Physics and Dance

Physics and Dance

Emily Coates and Sarah Demers

Yale
UNIVERSITY PRESS

New Haven and London

Published with assistance from the Alfred P. Sloan Foundation.

Yale University Press books may be purchased in quantity for educational, business, or promotional use. For information, please e-mail sales.press@yale.edu (U.S. office) or sales@yaleup.co.uk (U.K. office).

Printed in the United States of America.

Library of Congress Control Number: 2018933407

ISBN 978-0-300-19583-5 (hardcover : alk. paper)

A catalogue record for this book is available from the British Library.

This paper meets the requirements of ANSI/NISO Z39.48-1992 (Permanence of Paper).

10 9 8 7 6 5 4 3 2 1

Contents

Introduction

In certain respects, teaching physics and dance together seems obvious. Both physicists and dancers spend time imagining, enacting, modeling, and assessing motion. Physicists must identify and quantify the forces that act upon moving bodies. Dancers must move within a world of physical rules. Knowing more about these natural forces may help a dancer to understand dance technique more completely, and moving may help a physicist's imagination take flight.

Yet as much as they share a basic focus, physics and dance also differ in significant ways. The disciplines have very different approaches to methods and outcomes of research, modes of evaluation, and truth claims. Physicists and dancers acquire expertise through a wide range of practices—in varying degrees mathematical, experiential, embodied, and theoretical, depending on the line of inquiry in any given choreographic project or physics experiment. Their spatial scales do not always line up: in subatomic physics, the laws of motion change completely from those of the world available to human perception, in which choreographers' imaginations must live. Grappling with the differences between physics and dance can lead practitioners to new connections and unexpected revelations. The dance studio becomes a laboratory, and the organizing principles of planetary motion become a choreographic score.

Readers of this book will encounter a variety of interactions between physics and dance. In the first part of the book, we pair introductory topics in classical physics with basic principles of dance technique and relevant dance history. We organize these chapters according to terms common to both disciplines: gravity, force, motion, friction, momentum, and turning. In this part of the book, the connections between the disciplines are fairly direct. Physicists separate types of friction into static and kinetic, for instance, and deploy different coefficients and formulas to quantify the forces accordingly. Dancers recognize this same friction tactilely, as a force that impedes or enhances movement, giving a dance texture and meaning. The physics and dance pairings guide the reader through a physical encounter with the natural forces that physics describes while introducing some of the practices by which dance artists relate to those forces. We are interested in the multifaceted understanding that can develop from conjoining different ways of knowing the same thing.

1

In the second part of the book, we open up the inquiry to energy, space, and time—three broad concepts upon which the disciplines of physics and dance rest. In these chapters, we move more deeply into modern physics and choreographic research. Both have radically altered our perception of energy, space, and time, but they have done so using significantly different methods and under significantly different conditions. The realities predicted by modern physics are imperceptible to humans: we cannot travel at the speed of light or have a sensorial encounter with a particle. But choreographers can alter an audience's experience of time and space—in effect enacting relativistic conditions in live performance. Because the terms in which the research occurs diverge so dramatically, the connections between physics and dance in these chapters are less direct and not always obvious. As we weave through the various topics, we create certain links, through analogy or by tracing common threads. But we also intentionally give readers room to forge their own correlations and insights.

In addition to introducing readers to physics and dance, this book offers a primer in interdisciplinary research. Over the past seven years, we have collaborated in many different formats—from co-teaching to writing to creating art in different media. In that process, we have come to recognize that the nature of the connections we make between our disciplines can vary, from explicit to implicit, from the smoothly parallel to the seemingly far-fetched. Analogies might break down quickly or hold up surprisingly well. The point of interdisciplinary research is to assemble different ways of understanding the world, in order to ask questions whose answers might not be accessible through the individual disciplines alone.

Throughout the book we place equal emphasis on problem solving, movement exercises, and choreographic studies, without making hierarchical distinctions. These are practices that deepen and strengthen over time. They are also all forms of thinking—and the boundaries between aesthetic, mathematical, and embodied reasoning are blurrier than many would assume. Dance practice can be a form of quantitative research, and calculations can be a version of choreographic reasoning on the page. The consistent throughline in this book is movement.

Our approach departs from the usual teaching methods in both physics and dance. Physics textbooks typically present an assortment of hypothetical objects to explain concepts, including frictionless ramps, massless pulleys, ideal springs, and particles in boxes, as well as balls, cannons, athletes, coffee makers, and the occasional ballerina. Simplified shapes—such as a perfect sphere or cube—can give students a clearer introduction to new ideas. Concepts in modern physics, which are even further removed from lived experience, can require thought experiments: for Einstein's theory of special relativity, the teacher might use the light of a flashlight reflecting off a mirror on the ceiling of a moving train.

Imagining movement through idealized material objects allows for simpler calculations. But the prevalence of objects also places the reader at a remove—not as one who experiences the forces but as one who merely observes them. Frequently missing from standard physics textbooks is the recognition that the same forces that act on a box act upon the human body. A heightened awareness of the physics along with a myriad of political and cultural forces that shape human experience is precisely the research of dance.

Dance contests the idea that knowledge can be fixed or transmittable in books. Instead, the art form relies on a process of embodied transmission. Dance moves body to body, between individual people and among communities. The process is inherently active: to learn to dance, you need to move. In the passage from one body to the next, the embodied knowledge of dance, cognitively wired into muscle memories, slowly spreads.

Knowledge is codified in dance forms, which are styles or genres that preserve movement vocabularies and philosophies of motion. These details continually evolve as dance forms migrate across geographic distances and historical time periods. Take the many strains of Kathak, a form of Indian classical dance characterized by physical storytelling that originated centuries ago in northern India and is still performed today. Or in a very different example, George Balanchine's neoclassical ballet, in which he synthesized the cultural influences of nineteenth-century Imperial Russia and mid-twentieth-century America, specifically in New York City. The physical laws of our planet do not change. But the dancer's physical, psychological, and emotional relationship to those laws does change, depending on who is dancing, who is watching, what is being danced, and where the dance occurs.

Dance shows up in the archives through different forms of writing by artists and scholars. Systems of choreographic notation have cropped up over the centuries, often grounded in the biomechanical knowledge of the day. These systems always fall short of effectively preserving all the details of a dance in print. Dance has been documented on film and video— and these have also become a means of its transmission. Still, performance and dance scholars grapple with the ephemeral nature of the art form. Unlike standardized models in physics that hold up over many experiments, a dance will never occur the same way twice.

With such striking differences in the nature of physics and dance knowledge, what are the benefits of studying these disciplines together? The answer depends on your perspective. For science teachers, replacing boxes and pulleys with the physical experimentation honed by dance gives students different access to the forces studied—allowing them to move from what they know (moving through space) to what they may not know (the scientific analysis of movement). Analyzing the motion of dancing bodies requires a more sophisticated understanding of physics than analyzing the motion of boxes. For dancers, a more nuanced understanding of physical conditions can inform dance technique, infusing the dancer's thought patterns with a new awareness. Images drawn from physics can also inspire the kinesthetic imagination. For instance, imagining contact with the floor as an interaction between springs—which is the metaphor we use to explain

Newton's 3rd Law of Motion—will change the way the movement appears to the viewer and feels to the dancer.

Other creative and scientific benefits can also accrue. Using physics, a choreographer may gain new points of departure from which to launch into motion, and an expanded toolkit of prompts with which to experiment with energy, space, and time in choreographic composition. Meanwhile, the scientist gains a more multifaceted outlook on the basic concepts of energy, space, and time. Heightened choreographic imagination—a way of thinking about mass in time and space—can influence how people think through physics. The skills developed in each field are complementary. Choreographic research strengthens skills of observation, calculation, and problem solving. Quantitative reasoning cultivates the ability to engage with proportions and relationships, which are central to choreographic thought.

We do not always need to think about physics and dance simultaneously for the interdisciplinary dialogue to be in process. While the rewards of making direct "eureka!" connections are great, most of the time the work of putting these disciplines together requires going deeply into one field while pulling the other discipline along. Eventually, the focus will shift the other way. The inquiry occurs in motion, through a process of constantly changing perspective. Each discipline gives us lenses through which to view the other. The goal is to know which lenses to use when, and what to observe, in order to see something new. This work leads to something more subtle than a shriek of illumination: it quietly changes how we understand both physics and dance.

The physical conditions of our universe affect us whole bodily. The same natural forces that aid in the movements of balls and pulleys act upon us. We embody space and time, informed by cultural and political forces as much as by theories of relativity. The physics and dance inquiry in this book conjoins different ways of knowing our world to help us better understand how and why we move.

Part I: Principles of Movement

1. Gravity

Our study of gravity begins in motion. Start by lying on your back on the floor with your legs extended long, slightly wider than hip-width apart. Extend your arms out from your sides at a forty-five-degree angle. As you lie there, notice the environment you find yourself in. Are you lying on a soft and malleable surface or on hardwood or cement? Does the floor feel warm or cold? What sounds do you hear? How far away or nearby are those sounds? If you are in a building, are you on the ground floor or higher up, and how does this affect your senses? Notice the inescapable downward pull on your body.

Gravity glues us to the earth. When you get out of bed, you must push yourself upright. To walk around, you must push off against the surface of the planet. Without support—encompassing everything from your legs to the earth's crust—you would fall toward the center of the earth. A jump may give you a short-lived escape from contact with the planet, but even dancers who jump high cannot jump into space. This is a good thing, since the air you breathe, subject to the same gravitational attraction, also hugs the earth. You are attracted to the earth, and the earth is attracted to you, through the force of gravity.

The art of dance deals with gravity in many different ways. Some dance techniques are attempts to defy gravity, though its effects are ultimately inescapable. Other techniques opt to accept and give in to its force in order to produce movement. How we choose to dance with gravity shapes our sense of self and our worldview. Each dance form in turn expresses a particular vision for the interaction of humans and natural laws.

In this chapter, we will consider gravity both as a force that shapes human movement and as a natural law that we can model mathematically. We will help you build basic tools in both disciplines. You can improve your mathematical abilities with practice. And some of the most basic tools of dance are available to you simply by moving in the world.

Tuning in to Gravity

Returning to the movement exercise, begin to tune in to the forces that act upon you. As your body presses downward into the floor, the floor presses upward against you.

Check in with the various parts of your body. Lift one leg slightly off the floor. Hold it a moment and then gently set it down again. Lift the other leg ever so slightly off the floor, hold it there, and set it down. Your legs may shake slightly as your muscles hold them aloft. Now lift your arms and head a centimeter off the floor. Suspend them in the air and then lower them back down. How does the sensation of gravity change when you lift your limbs? What kinds of muscles must you flex in order to hold your leg or your arms a few centimeters above the floor? Your task is to remain fully alert, exploring in this isolated manner the forces exchanged between your body and the planet.

Now get up—but not as thoughtlessly as you might roll out of bed. Your movements should be more organized. Set a timer for eight minutes (the length of time is arbitrary, but setting a time constraint helps structure the movement). Rise to your feet at a constant speed, filling the entire eight minutes. You may find yourself shifting your weight slowly and purposefully in reaction to the effort that your muscles and skeletal frame exert to support you. The key is to cultivate a heightened attention to rising—an action you have done countless times.

Every so often, pause to monitor the sensation of gravity acting upon your body. As you work, release any muscles that might be tightening unnecessarily. Relax your head—let it drop toward the floor and sense its full weight. Relax your arms too, once you have risen to a point at which you no longer need them to push yourself up. With every pause, scan your muscles and release any unnecessary gripping; this will also permit you to feel the full weight of your limbs and torso. Feel the direction and quality of gravity's downward pull. Work slowly and deliberately.

This movement exercise gives you information about gravity. Through your actions, you have researched the effect of gravity on the mechanics of the human body. You have also paid attention to your own psychology and personal preferences in reaction to the natural force. Memories of prior physical training, old wounds or injuries, or other personal histories may affect the way you rise. We gain information through movement research: sometimes we can verbalize that information, and sometimes it remains a fleshly expression of physical and psychic awareness that cannot be articulated in words.

Gravity is powerful and omnipresent, like an extroverted relative who has an outsized ability to sway the family dynamic. Human beings recognize the force and calibrate their movements accordingly, because they have been living with it their entire lives, since floating in the womb. Most people tend not to shoot up like rockets when rising to their feet: their bodies learn how much force they must apply and in what direction to overcome gravity for the desired outcome or result. In dance, this knowledge is one aspect of muscle memory, the body's savvy ability to record certain actions under familiar conditions, including the way such actions feel and how to repeat them.

As much as dance artists rely on muscle memory to remember dances, they also look for ways to throw off old habits and reprogram what their bodies have learned in order to move in new ways. With subtle changes in approach, such as slowing down a movement, dancers can discover new information in the most familiar of actions. They defamiliarize everyday movements in order to transform them into an aesthetic representation, or art.

You can think of this first movement exercise as a *choreographic score*, in which the instructions organize the action. In this case, the score dictated that you rise up off the floor over a period of eight minutes, while pausing at your discretion to monitor your body's reactions. You had a time structure (eight minutes), a vector (up), and a directive (pause every so often). Add to these elements all the observations you made while executing this rise: the details and sensations that entered your awareness and the choices you made in response. Technique in dance requires sharpening your attention to the inner workings of your being and your environment, and to their interconnections.

The movement exercise that you just completed coordinates out of necessity with the physical conditions on our planet. If you were reading this book on the moon, which has a weaker gravitational field, you would weigh one-sixth of what you weigh on earth. Your movements would look and feel very different. But until humans are able to live somewhere other than the earth, all our dance forms must grapple with the laws of nature on our home planet, and choreographers must develop an attentive relationship to the gravitational attraction between the earth and the dancers. That relationship becomes the subject of the dance.

The Universal Law of Gravitation

Why are humans stuck to the planet? What is pulling us down? And how strong is this force? To understand the effects of gravity on our bodies through physics, we cannot think of ourselves in isolation. We need to acknowledge the presence of the entire mass of the earth.

As a result of centuries of meticulous observation, scientists have discovered that the gravitational force experienced by two masses depends on the amount of mass involved and the distance between the masses. The gravitational force that people on earth feel from the planet Jupiter is insignificant—even though the planet is three hundred times more massive than the earth—because Jupiter is millions of kilometers away. The gravitational force they feel from a dancer standing next to them is also insignificant, even though the person might be separated from them by mere centimeters, because human beings have relatively small masses on the gravitational scale. In order to experience a gravitational force strong enough to be felt by humans, the mass must be both large and close. The earth fulfills both of these requirements.

How do gravitational forces behave? We know about forces in the natural world that can repel, pushing objects apart. This is the case for the

electric force felt by two positive charges that are close to each other or two negative charges that are near each other. We also know about forces that can attract, like two electric charges that are oppositely charged, giving us a physics interpretation of the phrase "opposites attract." Gravitational forces are always attractive for a different reason: not because masses are opposite but because this appears to be the only way gravity functions.

Newton's Universal Law of Gravitation allows us to calculate the gravitational force (F_G) between ourselves and the earth, or between any two masses. To do this calculation we will label our two masses m and M. Our mass is the amount of stuff inside us—the amount of matter. The distance between the two masses is r. An important point to keep in mind is that the distance r is not between the surfaces of the two objects but between the *center of mass* of each object. This makes the distance between us and the earth not the width of the sole of our shoes but the distance between our center of mass and the center of the planet, which is approximately 6,500 kilometers (or 4,000 miles).

In order to figure out the strength of the gravitational force between the two masses, we also need a constant we label G. If we are measuring mass in kilograms and distances in meters, G is quite a tiny constant, with a value of 6.67×10^{-11} N m^2/kg^2, or 0.0000000000667 in units of newtons multiplied by square meters divided by square kilograms. Newtons (N) are the units of force. American readers will be more familiar with this unit's counterpart, the pound (lb), and 1 N is approximately equal to 0.22 lbs.

The equation for Newton's Universal Law of Gravitation is:

$$F_G = \frac{GMm}{r^2} \tag{1}$$

We can see from the equation that the force will increase when the masses get larger. And the force will also increase if we make the distance between their centers of mass smaller. The gravitational constant has the same value everywhere in the universe. This number gives us access to an innate characteristic of gravity.

Notice that though two masses are involved, we have calculated only one force. This is because each of the masses is pulled with the same attractive gravitational force. When we think about our gravitational attraction to the earth, we should not sell ourselves short in the relationship. In spite of our small masses, each of us is an equal partner with the earth. The strength of the earth's gravitational pull on us, what we call our weight, is equal to the pull that we exert on the earth. Because each of us has a relatively small mass compared to the planet's, the earth affects our motion, whereas we hardly affect the earth's motion at all. But it is perfectly accurate to think of the earth as our gravitational dance partner.

We can ask another important question: How is it that one mass is aware of and responds to the existence of another mass that is nearby? Scientists do not yet know. One theory posits that an as-yet undiscovered particle, the graviton, carries this information. Scientists still have more to learn about how gravity works. Gravity and general relativity are areas of active research, to which we shall return in the chapters on space and time.

With the Universal Law of Gravitation we can see that our mass is gravitationally attracted to every other mass in the universe, and all of those other masses are gravitationally attracted to us. That we are connected in this way to every mass in the universe is a little bit thrilling, and pushes the gravitational dance between ourselves and the earth out to a cosmic scale. But how does gravity influence our movement? How does it affect dance?

Balancing

Another basic movement exercise will help you deepen your understanding of gravity. Stand in parallel position: your feet should be approximately one foot's width apart and parallel to each other.

Close your eyes. Feel your shoulders over your hips. Turn your attention to the length of your spine, imagining its trajectory elongating up through the roof to the sky and down through the floor into the center of the earth. Relax your neck muscles by gently turning your head right and left. You have already become acquainted with the force of gravity while both lying down and rising. Now you are in a new position, standing. Your muscles and skeletal frame adjust to this new relationship to gravity. Many modern dance techniques incorporate this parallel position into their warm-ups.

Having focused your awareness on the forces at work in this stance, you are going to attempt a series of weight shifts. Keeping both feet on the floor at all times, shift your weight to your right and then return to center. Then sway to the left and return to center. Try leaning a few inches forward and backward. Try not to bend forward at the waist—think of yourself as a tower, adjusting in increments to small plate shifts under the earth. Be sure not to lock your knees. Unless you are playing a stiff-kneed character—a sailor in the 1940s musical *On the Town*, perhaps—no good dancing comes from locking your limbs.

As you perform this exercise, you will discover that your body makes minute muscular calculations to keep you from toppling over. The pressure in your feet will change according to where you distribute your weight. Your muscles tense accordingly, as well. The downward pull of gravity may seem to move around as you lean, becoming noticeable on your back, hips, head, or the side of your body, depending on the direction in which you lean. Think of your movements not as resisting but rather in conversation with natural forces.

The positions of the body through which dancers achieve balance reveal cultural, geographical, and historical attitudes. The exercise you just tried falls firmly within European-American modern dance traditions. Let's add another position to the research, for the subtle differences between dance forms and cultures can be astonishing seen through the lens of physics.

The idea of *sigi*, "sitting," in the Bamana language of West Africa, lies at the foundation of many traditional West African dances. The knees are bent,

as if the dancer is about to sit, with the feet still in parallel, approximately hip-width apart. The torso leans slightly forward, the spine long, as the pelvis shifts back to accommodate. When you try it, feel the difference between the completely upright torso of your first stance and this tipped pelvis and more forward torso. *Sigi* signifies an everyday action, sitting, even as it enables a readiness for the rhythmic play.[1] You may feel a delicate interplay between how far you tip forward with your torso and the counteraction of your pelvis, which helps you to maintain balance. The physics of gravity remains the same, but the dancer's position changes his or her relationship to those forces.

Dancers practice balance by following codified rules, which change according to the dance form. Is the torso upright or tilted? Are the arms far from or close to the body? How to hold the head? Dancers are not limited to balancing on two feet: sometimes they balance on their hands, head, shoulders, and toes. A spinning dancer faces a different challenge from that of a dancer who remains statically in place. A dancer spinning on her head has another challenge yet. And try to picture a dancer spinning on his head on a tightrope!

Balancing while standing is the first action involved in taking a step. Walking is nothing more than a series of off-balance forward weight shifts, a continuous series of falls due to gravity that allow people to move from one place to another. Try walking around the room. Walk on your heels. Walk on the balls on your feet. Walk leaning forward. Walk leaning back. Walk backward, opening your eyes to your environment. Feel the physics?

Center of Mass

The concept of center of mass in classical physics allows us to analyze how we are able to balance, to predict what will happen when we are off-balance, and to back up our theories with numbers. Center of mass is not only critical for a quantified understanding of balance—it is also needed to calculate the gravitational force acting on an object using Newton's Universal Law of Gravitation.

The simplest object in which to locate the center of mass is one in which the density (mass per unit volume) is constant throughout—that is, the center of the object itself would be the center of mass. As an example, a sphere's center of volume is at the center of the sphere. The same can be said for the location of the center of mass of a cube that is uniform in density.

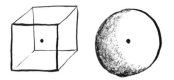

Of course, human bodies are neither simple geometric shapes nor uniform in density. In addition, human bodies are not rigid: the relative locations of their limbs, head, and torso can change. At this point we could throw up our hands (changing our centers of mass) and declare that the calculation of the center of mass of a person is too complicated to pursue. But fundamental concepts in our dance and physics interaction, like the conditions necessary for a person to balance, depend on the location of the body's center of mass.

In order to develop a general formula for the location of a person's center of mass, we need to develop a system to model the human body. But first let us think about a simpler case, in which a set of masses are all arranged on a line. We will impose an x-axis on this line of masses and designate a positive and negative direction. We also need to set a point as the origin, where $x = 0$. We will have positive values stretch out to the right and negative values stretch out to the left along our line.

The center of mass of a group of masses arranged on this line can be computed with the following formula:

$$x_{CM} = \frac{x_1 m_1 + x_2 m_2 + ... + x_n m_n}{m_1 + m_2 + ... + m_n} \tag{2}$$

where x_{CM} is the position of the center of mass along our x-axis. The values x_1, x_2, etc., correspond to the location of the masses m_1, m_2, etc. The formula to find the location of the center of mass adds a series of multiplications of an object's position by its mass in the numerator (the top half of the fraction).

14

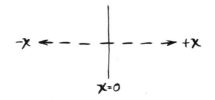

$-x \longleftarrow - - \quad - - \longrightarrow +x$

$x=0$

This gives us units of meters times kilograms, if we use the International System of Units (SI units). The sum of the objects' masses is the denominator (the bottom half of the fraction). Since the units in the denominator are kilograms, the final answer will be given in units of meters, and the results specify the center of mass position along the x-axis.

If we want to calculate the center of mass of a system that has masses in three dimensions (which is necessary for systems like people, who exist in three dimensions), we can repeat the calculation for the other two axes that make up our three-dimensional coordinate system:

$$y_{CM} = \frac{y_1 m_1 + y_2 m_2 + ... + y_n m_n}{m_1 + m_2 + ... + m_n} \tag{3}$$

$$z_{CM} = \frac{z_1 m_1 + z_2 m_2 + ... + z_n m_n}{m_1 + m_2 + ... + m_n} \tag{4}$$

It is convenient that the calculations in the various dimensions can happen independently of each other: if we constricted ourselves to moving along only one axis we could change the location of our center of mass along that axis without changing the location of our center of mass in the other dimensions.

Of course, the human body is a mass that is spread out over space, not chunks of mass condensed at points in space. We can make the calculation of a person simpler by handling each limb or section of the body independently. If we first calculate the centers of mass of the various parts of our body, we can then calculate the center of mass of our entire body by treating each body part as if all of the mass of that body part were located at its own center of mass. We could therefore have equations that look something like this:

$$x_{CM} = \frac{x_{CMLeftArm} m_{LeftArm} + x_{CMHead} m_{Head} + x_{CMTorso} m_{Torso} + ...}{m_{LeftArm} + m_{Head} + m_{Torso} + ...} \tag{5}$$

We can conceive of our bodies as a system of parts with masses and locations that can be combined in this formula.

15

How feasible is it to calculate these quantities for a person? Research can give us the average mass of men and women, and the average sizes and masses of parts of our bodies. If we are organized and careful, we could use these data to calculate the center of mass location for an average person, given a specific position. The calculations quickly become laborious, however, and stay relevant only for the period during which this "average" person remains motionless. But understanding how the calculations work can still give us insight into how our center of mass changes as we move.

Balancing Together

We can now calculate the center of mass of a system: What if that system is two people joined together as one? Another exercise can help this inquiry.

For this exercise, you will need a partner. Face each other and join hands by having your partner place her or his hands outward, palms up, while you lay your palms face down on them. From there, slide your grip up your partner's arm, so that you are gently holding each other's forearms. This point of contact should feel comfortable—don't grip too hard. In a moment you are going to be adding force.

Both of you should place your feet in a wider than parallel position and, if standing, bend at the knees, being careful to keep your knees in line with your feet, without rolling inward or outward. The torso is forward: picture the grounded position of an athlete ready to catch a ball or dart forward or backward. Your thigh muscles are engaged. You can think of this position as a deeper than usual knee bend—the stance allows you and your partner to modulate the tension between you, as if your knees were shock absorbers. The position allows each of you to sense both gravity and your partner.

Begin to pull away from each other—not too forcibly, just enough so that you would both be off balance if you let go. Maintain a curve in your lower backs. Keeping the stretched tension of opposition between you, allow first one partner and then the other to move in ways that change his or her center of mass. The other mover must make adjustments to accommodate those changes to prevent the system you are creating together from tumbling in one direction or another. One person explores the possibilities of shifting the position of his or her arms, feet, and legs, while the other must maintain the system's balance by shifting forward or back, deepening the bend in the knees, and so on. Take turns leading and accommodating the changes in center of mass. Once you feel comfortable with this first effort, explore different configurations, such as standing side by side and pulling away from each other holding on to only one arm. Try not to speak to your partner; instead communicate through physical negotiation.

This exercise allows you to research the conditions for balance between two systems that become one. People are different from inanimate objects, of course, and your movement research must also take into account the physical and social negotiation between you and your partner as one aspect of the information gained. But when we model this exercise using physics, we

temporarily set aside the fact of social interaction in order to focus on the interaction between your bodies and the earth.

Area of Support

When your center of mass—or that of the system consisting of you and your partner—extends beyond your base of support, you fall. But how can that limitation be quantified? If you stand with your feet directly underneath your hips about a foot's width apart, how large is the area of support?

If your feet are 25 centimeters long—a U.S. size 7 shoe for women or about a size 6 shoe for men—and a maximum width of 10 cm, we can make an approximation of the area of support as a rectangle that is 25 cm long and 30 cm wide if you stand with your feet in parallel. This gives a total area of

$$25 \text{ cm} \times 30 \text{ cm} = 750 \text{ cm}^2$$

If you stand with your center of mass directly over the center of area of support, you can lean 15 cm to the left and right, and 12.5 cm forward and backward before you begin to fall. If, however, you are working with a partner, clasping each other's forearms with your centers of mass separated by one meter when you both stand upright, the area of the base of support has been extended to roughly 1.250 m in length and 30 cm in width. Your combined area of support is then

$$1250 \text{ cm} \times 30 \text{ cm} = 37,500 \text{ cm}^2$$

Together, you and your partner are able to establish a larger area of support than either of you could individually. Assuming that your partner accommodates you, this allows you to lean considerably farther forward, backward, or sideways without losing your balance. Other sculptural forms that you and your partner might imagine might be possible to enact if you add a little physics analysis to aid in your discovery.

The Dance of the Future

The force of gravity on earth has remained constant for billions of years. But if we turn our lens from physics back to dance once again, we find that the nature of gravity—or more accurately, the dancer's relationship to gravity—has been more contested. In the development of American modern dance in the first half of the twentieth century, for example, gravity figured into polemical debates about power onstage.

In 1903, the dancer and choreographer Isadora Duncan gave a lecture that she later published as a manifesto titled "The Dance of the Future." Duncan's dance of the future stripped away what she saw as the artifice of classical ballet by tapping back into natural forces. As the cornerstone of her vision, Duncan argued for a new relationship to gravity:

> The dance should simply be, then, the natural gravitation of this will of the individual, which in the end is no more nor less than a human translation of the gravitation of the universe.[2]

In Duncan's view, the individual's will—his or her very impulse for moving—should align with the universal laws of gravitation. By working with, rather than against, gravity, her dance would channel a powerful cosmic force.[3]

Large ballet companies led by male directors dominated the Western concert dance world at the turn of the twentieth century. Duncan's manifesto was a direct refutation of classical ballet, whose trappings—from the training to the costumes to the patriarchal culture—she felt constrained dancers, especially ballerinas.

Her revolution played out in technical terms: while classical ballet technique asked dancers to pull *up*, Duncan's modern dance insisted that they pull *down*—toward the earth, rather than away from it. Being physicists as well as dance artists, we know that we cannot will our way into a new gravitational field—both ballet and modern dance ultimately deal with the same physical conditions. And the opposition that Duncan envisioned has become much less sharply defined in the one hundred years since she published her manifesto: ballet technique is more grounded and contemporary dance is more ethereal than she could have imagined. But the conceptual split that she identified between ballet and modern dance, based on the relationship of each to gravity, persists to this day.

With the deceptively simple act of performing earthbound movements that she had created herself, among other strategies, Duncan broke free of the restrictions imposed on women's bodies in performance. Gravity aided her symbolically, as much as physically, to tip the art of dance into a new sociopolitical age. For her innovations, historians have identified Duncan as a mother of modern dance.

Fast-forward fifty years to consider the work of the dancer-choreographer Pearl Primus, in which the dancer's relationship to gravity took on another political dimension. Primus was an anthropologist who incorporated her fieldwork into her dances. She was one in a line of anthropologically trained choreographers, including Zora Neale Hurston and Katherine Dunham, who shared an interest in synthesizing African diasporic forms with European-American modern dance. Offering another vision for the dance of the future that promoted Africanist cultures and aesthetics, these artists fought racism and championed the complexities of African American culture and history.[4]

Primus injected into her dances rhythm, fluidity, daringly high jumps, and fearless falls. Like Duncan, she danced with—not against—gravity, but in her case the gesture tied her to the dances that she had researched in the Gold Coast, Angola, Liberia, Senegal, and the Belgian Congo, among other countries on the African continent of the late 1940s.[5] She describes the earth in African dance as "an extension of the dancer's own feet, as if it were a stage of rubber from which he can bounce to the skies, as if it were a soft bed upon which he could roll and be protected."[6] Primus incorporated this Africanist vision of oneness with the earth into her choreography.

Study Primus performing her solo *Spirituals* in 1950 to witness the power of such a vision.[7] Primus leaps, legs in a V, arms open to the sky, then drops to the ground to execute a breathtaking series of forward rolls. Moving in a tight circle, she falls face-first onto the stage floor, rolls, and then quickly pushes herself back up to her knees—her center of mass low to the ground and her chest high, as if ascending from the deep. She does this over and over again. It's an image of redemption expressed in movement, and gravity is her guide.

Interdisciplinary Thinking

As you can see, physics and dance have different ways of understanding gravity and center of mass, and they do not always neatly coincide. In fact, their juxtaposition can be awkward, as we move from exploratory movement exercise and qualitative research to mathematical problem solving. This awkwardness is intrinsic to interdisciplinary thinking; it is also the source of interdisciplinary discovery. Remarkably, two bodies together, without using words or numbers, can figure out how to create a system that remains balanced. Also remarkable is that, armed with Newton's Universal Law of Gravitation, we can use the motion of our bodies to calculate the mass of the earth.

In this chapter you had an encounter with the process of synthesizing different disciplinary methodologies and ways of knowing to gain deeper understanding of each. As we proceed, we will ask you to keep an eye on questions of comparative knowledge acquisition: How is knowledge generated, what forms does it assume, and what can the combination of different ways of knowing tell us that we could not have found out through one discipline alone?

We are juxtaposing two different ways of understanding the forces of nature that affect us all. Physics gives us a framework within which to analyze and predict motion. Dance structures the perceptual experience that helps us understand the effects of this motion in our lives. The human body in interaction with the natural world is a highly complex system. As we have seen, we can heighten our awareness of natural forces through movement practices, even as we learn how to describe how these forces work using numbers on the page.

2. Force

The "duet with gravity" that you created through the previous chapter's exercises offered you a focused way of relating to natural forces. By slowly ascending from the ground, you acquired greater knowledge of the sensation of gravity upon your body. You also built knowledge of some of the ways that human anatomy can both resist and comply with that force. You experienced giving in to gravity and permitted the force to motivate your actions and shape your physical structure. You also cultivated an awareness of basic conditions for balance by creating positions—in physics terms, distributions of your mass—that helped prevent you from falling over. Dance builds upon these basic movement research experiments. Think of dance as a three-way interaction among natural forces, our awareness of those forces, and our physical imagination and capabilities in response to them.

You may have found that any previous physical training you have had—from flamenco to football—informed your experience of the gravity exercise. The information you acquire from practicing these physical techniques is one aspect of your embodied cultural knowledge, which is another form of "force" acting upon your body. Gravity is gravity, on earth, Mars, and beyond—but our physical, kinesthetic, and psychological relationships to these natural forces change according to the dance form we choose, and thereby the quality of the movement also changes.

Both natural and cultural forces shape how we move: the question is how to make those forces visible. Every dance practice embodies an entire cultural belief system—a cosmology of action and reaction in relation to the natural world. Physics probes those natural forces through theories, formulas, and diagrams. If we pull the scientific and cultural views together, we might be able to construct a more complete picture of human existence.

How do we measure motion? And what drives any form of human movement? The answer is force.

Newton's 1st Law of Motion

A dancer sitting on the floor must do some work to start moving and would need to do more to keep moving. This may make us think that our natural state is to be at rest. However, as we will see from Newton's 1st Law of Motion, the natural state of an object is more complex: it takes force to slow us down or speed us up, but not to keep us moving. Newton's 1st Law of Motion states:

A body at rest stays at rest and a body in motion stays in motion unless acted on by a net external force.

Forces are pushes or pulls. They are vector quantities: they have an amount— or magnitude—associated with them as well as a direction in which they are applied. To calculate our net, or total, external force, we need to add up all of the forces acting on us while taking into account the direction in which they are acting. Imagine you are standing between two people, with one person pushing you to the left and the other pushing you to the right. In order to calculate the total force acting on you, you need to take into account that these two forces are working against each other. A force of 10 newtons (N) to the left plus a force of 8 N to the right equals a force of 2 N to the left.

It might not seem to be a natural state of things for an object in motion to stay in motion because every time we start moving we have to work to keep moving, due to gravity and friction. One of the difficulties in understanding Newton's 1st Law is that we constantly interact with forces as we move. You may find that you need to change your intuitive sense of motion to conform to Newton's Laws.

As a start, imagine a place with fewer forces at play—outer space, for example. (You will realize as you read this book that physicists often send dancers into space. This is because the environment on the surface of the earth is much more complicated than the one in space due to the gravitational pull of the earth and frictional forces on earth's surface.)

Let's put you in a comfortable space suit with plenty of oxygen very far away from large masses, so that the force on you due to gravity is negligible. How would you begin to move from your position in space? You don't have many options. You need some kind of push or pull on your body in order to get started. And once you began moving in deep space there would be nothing to slow you down or stop your motion: no friction, no air resistance. In this context it seems perfectly clear that your natural state is to continue as you have been going, either at rest or in motion, unless you encounter external interference. This is exactly what is stated in Newton's 1st Law of Motion.

Is this also the natural state of things on the surface of the earth, where both gravity and friction come into play?

Let's continue our thought experiment and introduce a meaningful force of gravity but keep the forces due to friction minimal. We will put you in a skating rink with such smooth and slippery ice that you cannot get any traction. Gravity pulls you toward the center of the earth, but that force is exactly balanced by the ice pushing up on you, keeping you at a fixed distance from the earth's core. If you are at rest, you cannot start yourself moving. Your feet would just slip if you tried to run on the ice. Conversely, if some push or pull were to get you in motion, you would keep going (for a while) since there would be only the minimal force due to friction to slow you down. Again, these results are consistent with Newton's 1st Law.

Now try walking in socks on a dance floor (or in sneakers on the sidewalk). Frictional forces will require you to keep putting energy into maintaining your motion. This experience is what makes Newton's 1st Law of Motion so difficult to understand—it appears, due to these external forces, that our natural state is to come to rest. But remember, the continued effort to keep moving was not necessary in the ice rink or in outer space. External conditions can create frictional forces that interrupt an object's motion.

But before you conclude that friction acts only as an impediment to motion—it slows you down and requires you to work—keep in mind that it also gives you tremendous control over the direction you go and the speed of your motion. We will take a detailed look at friction in Chapter 4.

Newton's 2nd Law of Motion

Newton's 1st Law of Motion is really a subset of the 2nd Law. The 1st Law tells us what happens when there is no total force acting on an object—namely, that it will stay as it is, either in motion or at rest, whichever it has been. The 2nd Law allows us to calculate how the motion changes through acceleration due to the external forces. The law can be stated, first in words and then as a formula, this way:

The sum of external forces acting on an object equals the object's mass times its acceleration.

or

$$\sum(F) = ma \qquad (6)$$

where the symbol $\sum$ is a summation sign.

On the left side of the equation, the forces are added up. We must keep in mind that we need to keep track of their amounts as well as their directions because they are vector quantities. On the right side of the equation, the mass of an object m is multiplied by acceleration a, which is also a vector quantity. Imagine what happens to the acceleration when the amount of force changes but the mass remains unchanged. As the force increases, so will the acceleration. If instead the force stays the same but the mass increases, the resulting acceleration will be reduced.

This equation makes sense because it is more difficult to get an object with a lot of mass moving than an object without much mass. As m goes up, the value of a will decrease for a certain net force. It is also more difficult to slow down an object with a lot of mass that is in motion than it is a lower-mass object. This equation is incredibly useful because it lets us quantify the acceleration of an object under any set of forces.

In focusing on acceleration, Newton's 2nd Law of Motion can give us insight into one of the most universal dance movements: bending at the knees, which is often referred to as a *plié*.

The Plié

Bend your knees while standing, and you will appear to get shorter. But being "shorter" does not imply that you are less powerful: in fact, this movement will prepare you for any movement that might come next. Aided by your coordination, energy, and the type of movement that came before, a deceptively simple knee bend helps you muster the resources you need to dance. If we model this movement in physics terms, you first accelerate downward, and then you accelerate upward. The amount of the acceleration will differ, depending on the dance technique.

In classical ballet technique, this action is called a *plié*, a French word that means "bend" or "fold." The term has carried over to certain modern dance forms as well. Bending the knees lowers the body's center of mass, drops the pelvis, and allows for a more perceptible relationship to gravity. The plié allows for greater reactivity: a deep plié can mean the difference between having more force available to rev up and get going and trying to leap from a lock-kneed standstill.

The means by which a dancer accomplishes this simple knee bend vary dynamically depending on the dance technique. In certain West African dance forms, the knees bend swiftly, dropping the body's center of mass quickly. This plié acts less as a cushion than as a propulsive action through which to connect with the earth. From this lowered position, the dancer has control over rising, rhythms, and more. In contrast, in Bharatanatyam, one of the oldest forms of Indian classical dance, dancers frequently maintain a bend in the knees with the feet turned out. From this position, known as *ardhamandala*, the dancer drives the form's characteristic rhythmic footwork down into the floor. These various schools of thought express an entire ethics of how to move through, and relate to, the world.

Different schools of thought concerning the plié exist even within the same dance form—and individual teachers of the form sometimes differ the most. The Russian ballet choreographer George Balanchine, a founder of New York City Ballet, asked his dancers to accent the start of the plié and then slow the movement, as if applying shock absorbers. After reaching the bottom of the plié, which is very deep, the dancer would quickly return to a standing position, drawing the legs back together. The first count of the plié is thus a kind of bottomless resource, while the second count either returns the dancer to the starting position or results in surprising action: a pirouette, a jump in the air, a rise onto pointe.

Stanley Williams, a legendary teacher at Balanchine's School of American Ballet, took yet another approach. He used enigmatic imagery: "And you're *in*," he would say of a plié. Instead of thinking of the plié as a string of discrete opposites—moving the body down and up, bending the knees

24

out and in—Williams wanted circularity. Pulling up was part of lowering down, and pressing down was part of rising up. Likewise, the knees bending outward already contained the "in" of their straightening. Classical ballet comes with a set vocabulary of positions, which have been passed down and transformed from the fifteenth- and sixteenth-century European aristocracy to the present. Ultimately, Williams wanted to see no extraneous actions. His plié helped dancers transition seamlessly between positions, without the fidgeting that would have interfered with pure ballet form. Those who danced under Williams's tutelage felt that his plié possessed a kind of existential truth, and spent many hours attempting to perform his vision.

For our purposes, Stanley Williams's "in" might also be understood as "in the direction of gravity's pull," or inward toward the center of the planet. His vision of plié kept dancers from skittering on the surface. As with Bharatanatyam, West African dance, and many other dance forms, the plié and its variations are fundamental to dancing because all dancers need to figure out how to relate firmly to the ground upon which they move.

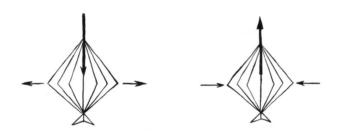

Down and Up

We left you with the equation $\sum(F) = ma$ and then veered off into the philosophy of bending at the knees and how various schools of thought in dance forge relationships to the earth. How do these relate to physics? In fact, we never left the physics behind, for every time you plié you are engaging with the phenomena described by Newton's 2nd Law of Motion.

To understand how $\sum(F) = ma$ is at work in the simple action of bending your knees, we will ask you to perform three variations on executing a plié in time. Keep in mind that with each version of the plié, you are playing with different accelerations along the vertical axis—Newton's 2nd Law should be whispering in your ear.

Stand with your feet in parallel position, one foot's width apart. Recall that your shoulders should be over your hips and your knees over your toes. Feel your spine elongate directly upward, as if hanging from hooks in the ceiling—a frequently used image in ballet and modern dance training. Be sure not to lock back in your legs: maintain a gentle give in the knees.

The ability to sense and react to the forces acting upon your body is largely contingent on your interaction with the force of gravity.

Now, we are going to organize your plié in time. You may recall that in the previous chapter we structured your rise from the floor in a unit of eight minutes at a constant rate. In this movement exercise, we are going to fine-tune that structure further and set the movement to *beats*, defined in dance by repeating, regular intervals of time. (Note that this is not how "beats" is typically defined in physics, where the word refers to the interaction between two frequencies, as occurs in tuning a musical instrument.) The beats can be slow or fast at your discretion, but the important thing to remember is that the beats give your actions a temporal framework—think of it as a support system for your movements that helps give them greater shape. Beats are the rudimentary building blocks of music, to which so much dance occurs.

In this exercise a metronome will be helpful. (Metronome apps are readily available online.) With your metronome ticking away, take two beats to bend your knees and two beats to rise up again. Be sure your knees are aligned over your toes when you bend, and your shoulders remain over your hips. Your head is held upright and your gaze steadily focuses straight out.

This is a plié in parallel, one of the most basic positions in modern dance technique. The evenness of these counts—two to go down, two to come up—have a metric regularity, like well-oiled pistons in a machine.

Now, for the second exercise, angle your toes slightly outward, keeping your heels together and your knees over your toes. This is first position in classical ballet. From this position, try a "demi"-plié, which means going as low as you can go while still keeping your heels on the floor, your spine elongated, and your pelvis tucked under. Use a different timing for this plié: accent the start of your descent on the first beat (loosely put, think of the accent as a "start quick then go slow" timing). By the end of the first beat, reach the full bottom of your plié. Then return to straight legs at the last second, by the end of the second beat. Try this a few times, to get the feel of it.

Finally, in this same turned-out position, try a third approach. Go to the bottom of your plié, maintain the bend in the knees and the turnout, and move forward around the space in which you are working. Advance heel first—that is, your heels should make contact with the floor first. Try changing directions, keeping the bend in your knees as you travel. Your thighs may burn, but no one said dancing was easy.

A good teacher will teach plié not as a rote exercise but as an act of discovery: of music and forces, self and world. Lessons on how to plié are best transmitted body to body by a live dance instructor, or even several live dance instructors, for each technique is a detailed language that dancers take years to master. To be sure, so much information is packed into any one dance step that a holistic description of the movement would far exceed the limits of the written page.

By performing three different types of plié, you have been researching how different technical approaches in dance operate and feel. You have also been researching Newton's 2nd Law of Motion, for the many technical details—such as timing, rhythm, and intention—lead to a unique acceleration in relation to gravity. We can deepen your research into the degrees and directions of the forces involved using what are called free body diagrams and force plates.

Free Body Diagrams

In order to analyze the three variations on a plié using Newton's laws, you can create a free body diagram. To make a free body diagram, first draw a representation of the object you are analyzing. Then indicate with arrows all the forces that act on the object. These are the forces you need to keep track of in order to do a calculation with Newton's 2nd Law. In this example, you are going to analyze a dancer doing a plié.

Keep in mind the distinction between the forces that your body is exerting and the forces that are acting on you. The forces your body is exerting include the force of your feet pressing down on the ground and that of your body pulling the earth toward it through gravitational attraction. The forces acting on you include the gravitational pull of the earth on your center of mass and the floor pushing up on your feet. Since your acceleration is determined only by the forces acting on you, include only those forces in the diagram.

Draw an arrow that indicates the direction in which each force is acting, and indicate the magnitude of the force, either with a numerical value or with a uniquely defined variable. Once you have included all forces acting on the body in your free body diagram, you can apply Newton's 2nd Law to calculate the resulting acceleration by adding the forces as vector quantities. In the sample free body diagram of a dancer in a plié that follows, we

indicate the force due to gravity with the label F_G and the force due to the ground with the label F_N.

A free body diagram is a powerful tool for analyzing the forces of nature acting upon a body. At the same time, the diagram has its limitations. It does not communicate history, for instance: it is a freeze-frame study, capturing the state of a body at one instant in time, without information about the motion that preceded that instant. A free body diagram of the body in midair, for example, would look the same regardless of whether the body was on its way up, at the height of a jump, or on its way down. In each instance, only the force due to gravity is at work. You can use the diagram to calculate the acceleration at that instant, which tells you how the object's current velocity is about to change. But the calculation will not tell you how fast the object is currently moving.

For the plié, a free body diagram of the moment when you begin your acceleration down will show a greater force due to gravity than that from the floor. In contrast, at the point when you begin to accelerate up a free body diagram will show that the stronger force acting on you comes from the ground. During the moments of constant velocity, whether you are moving up or down, the forces are perfectly balanced and there will be no acceleration.

A free body diagram leaves out other information as well, namely, the cultural contexts that inform human movement. Highlighting cultural forces within dance requires different modeling systems and theoretical frameworks, such as those found in the social sciences, humanities, and dance studies.

A Cultural Force Diagram

We used the free body diagram to map out the natural forces acting upon the human body in specific instances during a plié. What happens if we borrow this idea for the purposes of analyzing the cultural forces at work? We use

the idea of "culture" broadly here to describe both the codified rules inherent within a given dance form and the dancer's own personal movement history as it channels the social and political environments in which he or she lives.

Create a new diagram of your plié, using any of the variations on the plié that we have discussed. This time note any influences of movement training that you perceive to have had an effect on your experience of the plié. Do your knees long to turn in or turn out? Do you prefer to hold your spine elongated upright, angled, or stooped in relation to the floor? Why do your feet behave the way they do when you walk? Why do you hold your head steadily upright on your spine, or allow it to drop and react to gravity? What can you learn from how you hold your hands? Depending upon your prior physical training and social conditioning you may experience gaps—conflict, even—between the details of the dance technique and your body's preferred manner of moving.

Try to identify and spotlight details in the movement that relate to your own personal background and training. A wide range of physical training and environmental influences can count—from the types of spaces you usually occupy to your own internal sense of time, which is often culturally informed. (You might even include "impatience" as a cultural force inherited from the age of social media!) The goal is to unearth the movement cultures that influence how you move, for how you move shows you something about who you are.

Force Plates and the Plié

With the free body diagram and the cultural forces diagram, you have sketched out different forms of force acting on a dance movement. You can take your analysis further by adding a quantitative layer with the help of a force plate. A force plate enables you to measure forces that act on the plate's surface over time. Like a scale, a force plate continuously reads the force that presses upon it, but it is more precise in its readings than a typical bathroom scale, and it allows you to store the output on a computer for later analysis. If you stand still on a force plate, it will read your weight, which is your mass times the acceleration of gravity, since gravity is the force that is acting on you.

$$\sum(F) = ma = mg \tag{7}$$

where g is the acceleration due to gravity at the surface of the earth.

The reason you are not accelerating due to gravity as you stand still is that the floor is pushing back up at you with an equal and opposite force. As long as your center of mass has no motion with respect to the floor, the forces are equally balanced. How, then, do you lower your center of mass, as you do when you bend your knees in a plié? As you start to bend your knees, the force that the ground can exert on you is reduced. At least for a time, you will accelerate toward the floor.

Different variations on the plié will cause different accelerations and result in different force plate responses. You can imagine quickly accelerating to a constant velocity that you maintain for most of your plié. During the times when your body is moving with constant velocity, the forces of gravity pulling down on your body and the floor pushing up on your body are perfectly matched. This can occur even when you are moving toward the ground, as long as that motion occurs with zero acceleration and therefore constant velocity. When, however, you are changing your velocity, the forces acting on you are unbalanced, resulting in a non-zero net force and therefore an acceleration that can be calculated with Newton's 2nd Law. It is particularly instructive to try various plié techniques on a force plate so that you can clearly see the times when the acceleration is occurring and can map this information onto your motion. If you do not have access to a force plate, try the exercise on a bathroom scale. It will not be as sensitive as a force plate, but it should read something other than your weight when you are accelerating: a lower value than your weight when you are accelerating toward the floor and a higher value than your weight when you accelerate upward.

An example of what a plié might look like on a force plate is shown in the drawing that follows, labeled to indicate various components of the motion:

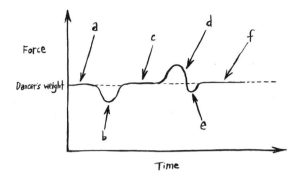

a : The dancer begins standing on the force plate, which reads his or her weight.

b : To begin the plié, the dancer starts to bend his or her knees and accelerate down.

c : There is a period of time when the dancer's center of mass descends with a constant velocity toward the floor.

d : At the bottom of the plié the dancer stops the downward motion and begins to go back up, which is a period of acceleration.

e : The dancer applies more force than needed to change the direction of motion. (This stage does not always occur.)

f : The dancer's center of mass moves at constant velocity upward, away from the floor.

Newton's 3rd Law of Motion

There is one more law of motion that will help to illuminate the forces at work in pushing up from a plié, as well as many other movements that a dancer undertakes: Newton's 3rd Law of Motion.

This final law refers to two objects exerting forces on each other, whether planets interacting through gravity or dancers moving in contact. This law can also explain the interaction between your feet and the floor, or your hand pressing on the wall of a dance studio or a chalkboard. The expression for Newton's 3rd Law of Motion, when objects A and B interact, is:

$$F_{AonB} = -F_{BonA} \qquad (8)$$

where F_{AonB} represents the force that object A exerts on object B and F_{BonA} represents the force that object B exerts on object A. The negative sign means that one force is acting in the opposite direction from the other. (Remember, forces are vectors, so they have both an amount and a direction.) Newton's 3rd Law tells us that these two forces are equal in magnitude and opposite in direction.

We have seen an example of equal and opposite forces with the Universal Law of Gravitation. People exert the same amount force on the earth as it exerts on them. The earth pulls people toward its center and people pull the earth toward their centers—the forces are in opposite directions. Objects interacting through gravitational forces are one example of Newton's 3rd Law behavior.

Initially, Newton's 3rd Law can be perplexing when it is applied to objects in contact. When you press your hand against a wall, the 3rd Law says that the wall "responds" with an equal and opposite force. Change the amount of pressure you apply to the wall, and the wall changes the amount of pressure it applies to you. How could the wall know how much force you are applying and respond immediately and appropriately? Where do these equal and opposite forces come from, and what constrains them to Newton's 3rd Law behavior?

When we think of dancers moving in the studio, we must remember that gravity is not the only source of force between objects. We need to also take into account the electrical forces that are involved between the atoms that make up the dancers and the floor. With regard to electrical forces, it is helpful to remember the common saying "opposites attract": two opposite charges attract each other. The strength of the force each exerts depends, like gravity, on how far apart they are from each other. But while the strength of gravity also depends upon the masses involved, the strength of the electrical force depends upon the charges involved. A positive charge and a negative charge will be attracted to each other with equal and opposite forces. When two positive charges or two negative charges move toward each other, they will repel each other, resulting in equal and opposite forces. In both scenarios Newton's 3rd Law applies.

How can we understand these forces when two objects are making contact, such as your hand against the wall or your feet on the floor? Although the details are more complicated than described here, imagine the surfaces of everything—you, the floor, the wall, your shoes—covered in springs, as a metaphor for the electrical forces involved. The atoms that make up your hand, the wall, and all matter have electrical charges in them. When you bring two atoms closer together, the interacting charges will make the surfaces repel each other. As you move two surfaces closer together, it is as if you were compressing springs, giving equal and opposite forces back on the objects in contact. When you push into the wall with your hand, the wall pushes back with equal and opposite force, giving you Newton's 3rd Law behavior because of the spring-like nature of the electrical forces involved. Even the gentlest contact with another dancer's fingertips will give the equal and opposite forces of Newton's 3rd Law.

How Do We Move?

A common confusion can arise with regard to Newton's 3rd Law: If every force has an equal and opposite partner, how can anything ever move? Push on a box, and the box pushes back equally. How do you slide it across the room? The answer is that only net *external* forces are counted when determining an object's motion. To calculate what a dancer's acceleration across a studio will be, we only need to know the forces that are acting *on* the dancer, not the forces that that dancer exerts on other surfaces. A free body diagram, which includes only the forces acting on the body in question, will help with this calculation.

If you are standing in the dance studio and you would like to begin to move, how would you accomplish this? As you stand on a floor you have at least two forces acting on you: the force of gravity due to the interaction between your mass and the earth's, and the force of the floor on you. If you stand at rest, with zero acceleration, these forces are perfectly balanced. The "springs" on the bottom of your feet and the surface of the floor have compressed.

So how do you get the unbalanced force you need in order to begin to move? One option would be to have someone give you a push (carefully). That could start you moving. But what if you wanted to start moving on your own?

All you need is contact with something. In this instance, you are in contact with the floor. If you push on the floor, the floor has no choice but to push back on you with an equal and opposite force. If you were to draw a free body diagram of this scenario, keeping track only of the forces acting on you, you could see how your pushing resulted in a net force that would drive you in a direction opposite to the one you applied to the floor. If you want to move forward, push backward against the floor. If you want to move to the left, push to the right; and so on. The more force that you successfully apply, the greater your acceleration in the opposite direction.

In one simple formula, Newton's 3rd Law of Motion illuminates the mechanics of human movement. A surface you push against actually pushes back on you, which makes possible a full spectrum of dance forms. Think of this transaction as a continuous dance with your environment: you are never dancing alone.

Everyday Action

The relationship between forces described in Newton's 3rd Law enables us to perform all of our daily actions—sitting, standing, walking, running, skipping, jumping. Whereas physicists can predict the average forces involved in each of these actions, suggesting a certain consistency in their value and interpretation, how a choreographer frames those movements can alter their meaning entirely.

Consider the action of scrubbing the kitchen floor on hands and knees, which the dancer-choreographer Blondell Cummings incorporated into her seminal 1981 solo, *Chicken Soup*. As Cummings choreographed the dance, the performer alternates between repetitive, rhythmic scrubbing—her long arms pushing the brush away and back toward her body, away and return—and up-on-her-feet dancing, which is wiggly, energetic, seemingly unbounded. Cummings's composition alternates formally between the recognizable action of scrubbing and the more ambiguous meanings of her upright dance. As the solo progresses, she dance-ifies the movements involved in cooking, too: through rhythm, repetition, and expansive execution, they become choreographic. The set minimalistically evokes a kitchen, with a table and chair. Cummings wanted to create a solo about women and food, which spoke not only of her own African American family background but to a universal domestic experience for women across cultures.

What are the forces involved in Cummings's movements? From the perspective of physics, we can think about her actions quantitatively. We know from Newton's 1st Law of Motion that she needed a force to get going. There are biomechanics involved here—food becomes energy for the muscles that press into the floor to create Cummings's rocking, scrubbing motion. Newton's 3rd Law allows us to calculate the force she exerts into the floor and the force the floor exerts back on her. We could also measure the angle of those forces, as we shall see in Chapter 4.

Apply Newton's 2nd Law to calculate the force of her arm muscle and the force from the floor upon the scrub brush, and you can calculate the brush's acceleration. However, as you might notice should you do a comparative study of floor scrubbing across contexts, her force and engagement with the brush possess an unusually dynamic intensity, for she is dancing the action. A snapshot $F = ma$ would not necessarily pick up on the waxes and wanes in force that constitute the rhythmical structure by which she turns her scrubbing into dance.

Cultural forces are involved, as well. This is a dance about the socially determined gender roles prescribed throughout American history, represented in the domestic space of a kitchen and Cummings's tasks. Also salient are the critical interpretations that viewers confer upon race. Although Cummings intended the dance as a universal statement about womanhood, because she was a black woman the solo has frequently been interpreted as a black protest piece.

When Is Walking More Than Walking?

Everyday action in concert dance takes on different meanings depending on the artist, time period, and audience. Blondell Cummings worked in an aesthetic lineage that reached back to the early 1900s, when Isadora Duncan was reacting sharply against what she believed was the harsh artifice of classical ballet training on the body and designing a way of dancing she deemed more natural. Duncan's movement was not totally extracted from precivilized existence—her aesthetic influences included the philosophy of

Nietzsche and classical Greek sculpture. But she nonetheless opened the floodgates to a new way of dancing before the public—solo, as a white woman—partly by including movements that many in the audience could have performed.

In Weimar Germany in the 1920s, the choreographer and theorist Rudolf von Laban not only incorporated pedestrian movements, he also opened dance to the untrained when he created massive movement choirs with hundreds of participants. Everyday action became large-scale dance spectacle in his hands.

In New York City during the 1960s, an adventurous group of young choreographers pushed the type and range of actions that could be included in dance farther, partly by inventing new choreographic structures to organize those movements. They created rigorous spatial and temporal scores to contain everything from walking and running to moving mattresses and eating apples. The group presented their work under the collective Judson Dance Theater and have since come to be known as pioneers of postmodern dance.

Cummings emerged as a performer in the late 1960s out of Judson circles of influence. She developed her own unique spin on composing with everyday action: whereas Judson artists *de*contextualized everyday behaviors in their compositions, Cummings returned them to their original context by alluding to their settings—as with the spare set that evoked a familiar kitchen—thereby charging those actions with a social commentary all her cwn.

Using pedestrian movements, dance artists have challenged classical ballet's elitist domination in dance, brought high art to the masses, democratized the space of public performance, and drawn attention to marginalized experiences. When the movements onstage look more like everyday life than virtuosic technical feats, hierarchies begin to topple.

If walking is presented as walking—without expressive flair or kinesthetic flourish—what pushes it into the realm of art? One way to answer this question is to consider the artist's compositional choices: the temporal structure, spatial organization, and movement vocabulary all impact the viewer's perception of any movement, whether spectacular or quotidian. By subtly arranging a walking pattern, or juxtaposing movements against each other—the action of scrubbing against interpretive dancing, for instance, in the case of Cummings's formal innovation—the choreographer can render something that is familiar unfamiliar. Through this defamiliarization, we can be jolted into seeing our world afresh. To quote the choreographer Yvonne Rainer quoting the composer John Cage (who was himself repurposing a verse in the book of Ecclesiastes), there is nothing new under the sun; there are only new ways of organizing it.[8]

3. Motion

Most people get through the day with a very narrow range of foot activities: their feet usually assist with standing and transporting them from here to there, often by walking or running. But feet have so much more to say than this limited repertoire suggests. Consider the many different ways your feet can move. In dance, feet can slide, slap, tap, stomp, flex, flick, relax, and point. They can help the dancer leap upward and outward, and then cushion the descent when he or she lands.

When it comes to leaping in physics, Newton's 3rd Law of Motion establishes that as people push off the planet to get moving, the planet pushes back at them. If they manage to push hard enough, the planet can launch them into the air—at least for a brief period of time. A new set of equations enables us to access the symmetry that lurks in the phenomenon of jumping. Combined with an investigation into the nitty-gritty details of how dancers jump, these equations can help illuminate how dance achieves its expressive power.

Physicists typically set up hypothetical situations with objects to explain how to calculate the variables involved in projectile motion. In this chapter, we replace scenarios that use inanimate objects with dancers and their physical techniques. One way to understand physics concepts is through the motion of the human body—and few people have greater agency over the physical forces acting on their bodies than dancers. A dancer's knowledge will affect the quality, height, and timing of the jump, even as the physics tells us how high, how far, and how long the dancer can manage to fly.

Dancing Feet

Dancers use their feet in specialized ways: to slam down or roll through, to tread softly or consume space, to produce musicality and rhythm, to convey character. In dance, feet speak.

Acquiring sophisticated knowledge of footwork takes practice over time. Depending on the dance form, a dancer's feet develop different habits. Ballet dancers tend to point their toes any time their feet leave the floor. Other dancers use what we might call a "relaxed point," which is somewhere between pointing and actively flexing at the ankle. Tap dancers strategically relax at the ankle, while classical Indian dancers flex at the ankle. Ballroom dancers are judged by the clarity of the patterns their feet make on the floor.

Postmodern choreographers mix the ordinary footwork of walking with extraordinary phrases of dancerly movement. In Memphis jookin', dancers mix modes as well, even twirling on the tips of their sneakers. How dancers use their feet to move reveals both their own personal history and the larger cultural history of the dance form.

But we are not concerned solely with how a dancer gets from here to there horizontally—that, we know, is assisted by Newton's 3rd Law of Motion. We are heading in this chapter toward another dimension of dance: leaving the ground; not only by jumping directly upward, but also by leaping up and out. What resources do dancers use to soar?

Taking Off and Landing

Jumping in dance usually involves three components: the feet, the knees, and the arms. The feet and the knees, in plié, provide the engine for flight. Dancers push off through their feet. And just as a spring coils and releases, the plié that precedes a jump dictates its height. The plié also cushions the descent. The arms are important as well: note the strategic use of a long jumper's arms in flight. Dancers similarly swing their arms so as to add greater height and distance to a jump.

We have been thinking about feet, so we will focus there. Musicality, timing, and ability to react all lie in a dancer's use of the feet. Dancers use their feet to jump in a variety of ways, depending on the dance form. In one school of thought, dancers "roll through" their feet on the way up from the plié and into the air, and then reverse the sequence on the way down. "Rolling through" the foot means that once the legs are nearly straight out of the plié, first the heel leaves the floor, then the ball of the foot presses through, and then the toes give the final push upward. On the way back down, the dancer reverses this process by way of contact with the floor: the toes meet the floor, then the balls of the feet, then the heels.

Perhaps counterintuitively, jumping technique often focuses on the landing, more than on achieving great heights. Rolling through the feet gives a dancer tremendous control, especially in the descent to the earth. Imagine a cat leaping off a counter to the ground: it does not crash down awkwardly and struggle to recover. Instead, it heads straight into its next action. Dancers practice rolling through their feet over and over again in order to build the strength required to land as gracefully as a cat.

Other schools of thought might include landing on the balls of the feet, or on the full flat foot. Sometimes an ascent has less to do with achieving great height than with coming right back down to execute grounded rhythms that require deliberate weightiness into the floor. The landing is frequently more important than the jump itself—adding yet another dimension to the dancer's duet with gravity.

Think for a moment about how many different ways a dancer with the full use of both feet has to take off and land:

Taking off from two feet / Landing on two feet
Taking off from two feet / Landing on the left foot
Taking off from two feet / Landing on the right foot

Taking off from the left foot / Landing on two feet
Taking off from the left foot / Landing on the left foot
Taking off from the left foot / Landing on the right foot

Taking off from the right foot / Landing on two feet
Taking off from the right foot / Landing on the left foot
Taking off from the right foot / Landing on the right foot

That's nine ways to take off and land using the feet. The dancer's anatomical structure is to a choreographer what metrical beats are to a poet. When writing sonnets, poets make art out of a set number of beats per line. Similarly, a choreographer makes art out of limits imposed by the human form. And we have not yet even covered other ways of taking off and landing, which might involve the hands, back, shoulders, and even head. Choreographers can multiply the potential of the human body in ways that many people have never imagined.

To understand the basic mechanics of landing, try this exercise. Find parallel position. Working gently, perform four jumps straight in a row, practicing rolling through the feet as described above. Setting a metronome will help keep you on a beat. Jumping a centimeter or two off the floor works just fine—no need to hit the ceiling. For dancers of different physical abilities, this exercise can also be practiced with the hands against the thighs or a wall, to help them investigate the effects of pressing through to take off and cushioning the descent.

Projectile Motion

Once you have caught your breath and recovered from your investigation into jumping, physics has more surprises for you.

One of the most powerful ideas in the analysis of motion is that action in one axis can be analyzed independently of action along another axis. First you must situate the motion—such as a dancer leaping through the air—within a coordinate system. You can then analyze the vertical component of the leap, along the y-axis, separately from the horizontal component, along the x-axis. The link between motion in one axis and motion in the other is that time is flowing at the same rate in each.

Let's first introduce the variables you will be manipulating as you work with modeling motion in physics. You will need variables that allow you to keep track of a mover's position at any given point in time. So the motion must first be put within the context of a coordinate system. You can keep track of the position of the center of mass of a dancer, or a foot or hand, using the coordinates x, y, and z. This assumes that you have declared the directions in which these axes are aligned and the point at which $x = y = z = 0$. At any given moment, you can figure out the position of something by knowing the value of the three coordinates.

To construct the axes, align the direction of gravity with the y-axis, as is the convention. The direction toward the center of the earth (down) dictates the negative y direction. Positive y therefore points directly up toward the sky.

If your studio is a square or rectangle, you might find it simplest to let the x- and z-axes be aligned with the walls. Or the x-axis could be aligned with the particular direction that a dancer is moving in the motion that you are analyzing. Set the $x = y = z = 0$ point at a location that will make your calculations simplest. You may find that the floor provides the most convenient $y = 0$ position. Or it might be simplest to set $y = 0$ at the initial or final position of a dancer's center of mass. It takes practice solving problems to get better at making these decisions. As long as you define your coordinate system consistently within a problem, your calculations should give you correct answers.

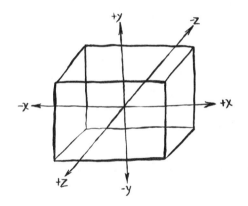

In addition to predicting or keeping track of the position of a person or thing, we also want to understand the rate at which the position is changing. This is the *velocity*. While a position can be given in units of distance, like meters, velocity deals with the change of that position as a function of time, so has units of meters over seconds (m/s). You can use v_x, v_y, and v_z to keep track of the velocity in the x-, y-, and z-axes, respectively.

Finally, we also care about how the velocity of the object changes as a function of time. This is the object's acceleration. Since acceleration represents velocity over time, it is calculated in meters per second per second, which gives m/s^2. We use a_x, a_y, and a_z to keep track of the acceleration in the x-, y-, and z-axes, respectively.

Let's start with a bite-sized piece of motion, but limit the analysis in a way that still allows you to do something interesting—not a trivial task, given how complicated motion in dance is. A set of equations is available that can be used in cases where acceleration is constant. If you constrain yourself to analyzing movements when dancers are in the air, the only relevant force acting on them can be gravity, which provides a constant acceleration in negative y in our coordinate system. Assume that the dancers are close to the earth's surface, where the constant acceleration due to gravity is denoted by the variable g, with a value of 9.8 m/s^2 in negative y.

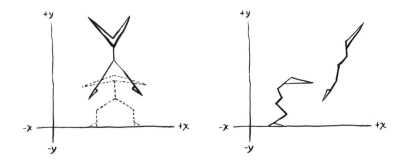

Since the dancer is only under the influence of the constant force of gravity in this simplified picture, he or she has two possibilities for jumping: vertically directly up and down, with motion constrained in y, or sailing outward, with a horizontal as well as vertical component to the jump. To simplify the task, arrange the coordinate system so that the x-axis is aligned with any horizontal motion that occurs; then you can work with just the x and y coordinates and not worry about z.

Remember that in the chapter on force we had only the force diagram and information about the acceleration of the object at one particular instant. Now bring in the past or the future as appropriate, by including in your model where the jump came from and where it is going. In other words, you are making a new model in which you will be able to look at changes in positions and velocities over time. In any given problem, you will need an initial set of conditions, labeled by i, and a final set of conditions, labeled by f.

You must set a start time that defines the initial conditions and an ending time that defines the final conditions. Choose these two times to correspond to the instants that you want to analyze. You then have the following variables for the start of the movement:

variable	description
x_i	initial position along the x-axis
y_i	initial position along the y-axis
v_{xi}	initial velocity in the x direction
v_{yi}	initial velocity in the y direction
t	time between the initial and final shapshots

42

At the end of the movement, each of the position and velocity variables is repeated, but with the initial i swapped out with final f:

variable	description
x_f	final position along the x-axis
y_f	final position along the y-axis
v_{xf}	final velocity in the x direction
v_{yf}	final velocity in the y direction

Last but not least, we must recognize the effect of the constant acceleration. That acceleration will be equal to g, which is expressed as $a = g = 9.8$ m/s^2. This is the acceleration due to the force of gravity acting on a mass near the surface of the earth, and it only acts along the y-axis. There is no acceleration along the x-axis.

Equations of Constant Acceleration

The framework that you are building is referred to as projectile motion in physics. Its equations are used for cannonballs, boxes that you toss in the air, missiles, and more. If you don't mind tossing yourself into the air—jumping—then you can also be considered a projectile. Your center of mass will follow the equations for projectile motion.

We'll start by introducing all the relevant equations. Then you can work with them.

$$x_f = x_i + v_{xi}t + \frac{1}{2}a_x t^2 \tag{9}$$

$$a_x = \frac{v_{xf} - v_{xi}}{t} \tag{10}$$

$$v_{xf}^2 = v_{xi}^2 + 2a_x(x_f - x_i) \tag{11}$$

In equation 9, we see the relationship between position, velocity, and acceleration. We can read through the equation as we would read a sentence: The final position along an axis (x_f) can be calculated by first taking into account where you start (x_i). In order to predict where you will end up after a given period of time, you also need to take into account how fast you were moving when the clock was started. We accomplish this with the term ($v_{xi}t$), which takes your initial speed and multiplies it by the time you will be traveling. You would now be done if your velocity never changed. We accommodate changes in velocity—acceleration—and the duration of motion over which the acceleration acts with the term $\frac{1}{2}a_x t^2$.

This equation will only work if your acceleration is constant over the time that you consider t. If the situation were more complicated you would need a more complicated equation.

Before we turn to a calculation, take another pass through the equation, this time with a different focus: units. You will see that on the left the variable (x_f) has units that match the distance scale, meters. You should therefore expect every term on the right-hand side of the equation to have units of meters too, or the equals sign will have been a mistake.

The first term, x_i, is in units of meters. The second term, $v_{xi}t$, consists of a velocity multiplied by a time. With velocity in units of meters per second (m/s) and time in units of seconds, you can see that the impact of multiplying velocity by time cancels out the seconds, leaving only meters. So far so good! The third term, $\frac{1}{2}a_x t^2$, begins with a unitless number $(\frac{1}{2})$ that you can ignore for the purposes of checking unit consistency. What remains is an acceleration (m/s^2) multiplied by time squared (s^2), and, again, here the seconds cancel out, leaving only meters.

Equation 10 tells us that the acceleration is equal to the difference between the final and initial velocities, divided by the time. A quick glance at units here shows you that to achieve the units of acceleration (m/s^2) on the right side of the equation, take the units of velocity (m/s) and include an additional time term (in seconds) in the denominator.

Equation 11 allows you to compare the final velocity with the initial velocity. The two will be equal only if the acceleration is 0. Otherwise the initial velocity squared must be adjusted by a factor of $2a$ times the total distance over which the acceleration is applied (x_f - x_i). We leave it as an exercise to you, the reader, to check that we have consistent units on the left and right sides of the equals sign.

You may be wondering why we need so many equations when each seems to be relating similar variables in similar contexts. One reason is for convenience. The first equation can be used regardless of whether you know, or care about, the final velocity v_f. The second equation, on the other hand, does not rely on the positions or total distance traveled. And the final equation is written in a way that does not use time. When doing a problem with projectile motion, it is important to first understand which variables you know and which one(s) you must calculate, so that you can choose the most appropriate equation to work with.

Jumping Up and Out

If we substitute dancers for objects in our physics analysis, we need to take into account the techniques with which dancers jump through space. More specifically, we need to look at how dancers jump not only up, but *up and out*.

At this point, you have acquired an arsenal of physics knowledge that you can translate directly into dance technique. Newton's 3rd Law of Motion tells us that the direction of the jumper's force into the ground directly affects the direction and angle of the jump. Push straight down into the floor, and the floor pushes you up. Push at an angle, and the floor pushes you outward at that angle and you will travel.

Dancers can modulate the same jump to different effects, by changing the angle of their force into the floor. More loft in the air requires a sharper angle downward. Covering more distance with any single jump requires pressing into the floor at a shallower angle. The same principles regarding the use of the feet and the plié apply.

Swinging the arms down and then up can add an extra element of force. But the reason why this is effective is counterintuitive. The arms do not augment the momentum—an image that many dance teachers use to teach jumps—so much as generate extra force. Swinging your arms upward in fact gives you greater force into the floor.

You can check this out by swinging your arms forward and back while standing on a scale. The scale will read above or below your weight at different points in the motion, reflecting the higher and lower degrees of force into the floor, depending on the acceleration of your arms at any given moment. Paying attention to the increased pressure between your feet and the floor as you swing your arms is another way to research this phenomenon. Interestingly, maintaining a constant velocity with the arms does not result in the same force pressing into the floor. It is only when the arms are accelerating that the force downward is amplified.

To test all of these ideas, try *skipping*—a favorite action of adventurous children and choreographers. Skipping is walking and jumping on a rhythm: "step, JUMP, step, JUMP, step, JUMP" like so. The jump can either take you directly upward or help you travel a longer distance.

Skipping is an ideal action with which to experiment with pushing into the floor at different angles. You should find that if you push straight down into the ground before the jump, your skip will be quite high. If you push into the ground at an angle, however, you can travel several feet during your jump before you come down—creating a space-eating effect.

If you map a coordinate system onto these actions, you can note that the first approach (jumping straight up) moves some distance along the y-axis but less along the x-axis. And if you map the second approach (pushing into the ground at an angle in order to travel), you will discover that the skip has moved some distance along the x-axis.

Remember that pedestrian movements, thoughtfully composed in time and space, can be found throughout many dance forms and choreographic works. We could impose a three-beat time structure on skipping, for instance, consisting of one beat to walk, two beats in the air. Or we could change the pattern of the footsteps to "run, run, JUMP, run, run, JUMP," giving the skip a different impulse. How high you hike your knees in the air with each skip creates character.

Whichever way you try to skip, remember that pushing *straight down* will help you go *up*, and pushing down *at an angle* will help you travel *out*. This basic concept lies at the heart of most jumps in dance.

Calculating a Jump

Let's test our equations of projectile motion on a vertical jump—a jump straight up and down that we can model along the y-axis alone. The problem is to calculate how fast you must be going when you leave the ground in order to jump to a height of 0.2 m above the ground. To be very precise, we will ask that your center of mass moves 0.2 m up along the y-axis. Because your legs might be bent and your body will not be rigid throughout the jump, it can be tricky to think about the relationship between your feet and the ground. It is simpler to keep track of the motion of your center of mass than to take these details into account.

First go through the mechanics of the jump with each of the variables in our equations. We list only the y-axis variables here, since you are constraining yourself to motion up and down along the y-axis:

variable	description
y_i	initial position along the y-axis
v_{yi}	initial velocity in the y direction
t	time between the initial snapshot and final snapshot
y_f	final position along the y-axis
v_{yf}	final velocity in the y direction
a_y	acceleration, which is 9.8 m/s^2, in the negative y direction

We are faced with a choice about where to set $y = 0$. We will set $y = 0$ as the y location of your center of mass when you are standing on the floor, just as you are about to leave the ground. Now we have to make a choice about our initial and final snapshots. Let us set our initial time as the instant just

46

as your feet are leaving the floor. Our final snapshot in time can be set at the height of your jump, when $y_f = 0.2$ m.

We know that the acceleration will be $a_y = -9.8$ m/s^2 for the entire jump. Gravity is always acting to slow down your upward motion or speed up your downward motion. We can rewrite the list of variables, filling in the details for the ones we know and leaving the others as question marks thus:

$$y_i = 0.0 \text{ m}$$
$$v_{yi} = ?$$
$$t = ?$$
$$y_f = 0.2 \text{ m}$$
$$v_{yf} = 0 \text{ m/s (at the height of your jump)}$$
$$a_y = -9.8 \text{ m/s}^2$$

There are two unknown variables in the above list: the initial velocity in the y direction (v_{yi}) and the amount of time t it takes you to reach the height of your jump. Which do we care about? In this problem we are tasked with calculating how fast you must be going when you leave the ground in order to achieve the height of 0.2 m. That means we want to calculate the v_{yi}. Look back at the equations for projectile motion (equations 9–11). Is there one that will allow you to calculate v_{yi} given the information we have?

Equation 11 will work if we convert it to y coordinates instead of x coordinates:

$$v_{yf}^2 = v_{yi}^2 + 2a_y(y_f - y_i) \tag{12}$$

Plugging in the values that we know, we have:

$$(0 \text{ m/s})^2 = v_{yi}^2 + (2)(-9.8 \text{ m/s}^2)(0.2 \text{ m} - 0.0 \text{ m}) \tag{13}$$

Simplifying this equation, we get:

$$0 = v_{yi}^2 + (-3.92 \text{ m}^2/\text{s}^2) \tag{14}$$

We can then subtract v_{yi}^2 from both sides of the equation, giving:

$$-v_{yi}^2 = -3.92 \text{ m}^2/\text{s}^2 \tag{15}$$

The two negative signs cancel each other, so we can take the square root of 3.92 m^2/s^2 to get our final answer of a velocity of approximately 2 m/s. If you wanted to jump higher than 0.2 m, you would need to be going faster than this value. If you wanted a lower jump, you would need to be going slower when you left the ground.

For completeness, let's list the full set of known variables again:

$$y_i = 0.0 \text{ m}$$
$$v_{yi} = 2.0 \text{ m/s}$$
$$t = ?$$
$$y_f = 0.2 \text{ m}$$
$$v_{yf} = 0 \text{ m/s}$$
$$a_y = -9.8 \text{ m/s}^2$$

Can we now calculate the amount of time it takes you to reach the height of your jump? Look back at the projectile motion equations once again: we could use both equation 9 and equation 10 for this, but the latter will give a simpler calculation because time is not squared. Let's take equation 10 and solve for time before replacing the variables with numbers. To do this, we multiply both sides of the equation by time and divide both sides of the equation by a_y, which will result in this equation:

$$t = \frac{v_{yf} - v_{yi}}{a_y} \tag{16}$$

Now, plugging in the numbers in place of the variables, we have:

$$t = \frac{0.0 \text{ m/s} - 2.0 \text{ m/s}}{-9.8 \text{ m/s}^2} = \frac{-2.0 \text{ m}}{-9.8 \text{ m/s}^2} \tag{17}$$

Again, the negative signs cancel each other. The time therefore equals:

$$t = \frac{2.0 \text{ m/s}}{9.8 \text{ m/s}^2} = 0.2 \text{ s} \tag{18}$$

You can go back over each of these examples and note that the units do, in fact, remain consistent on both sides of the equals signs throughout. Keep careful track of units while you work in order to catch mistakes that can creep in when you do these problems.

There is a neat symmetry to a jump through the air that you can exploit when you do your calculations. The first half of your jump consists of your leaving the ground at your maximum velocity for the jump. Gravity then works to slow you to a stop at the height of your jump. From this point, gravity pulls you back down until you are going the same speed at which you took off, but in the opposite direction, as you land. You spend as much time going up as you do going down. The total time that you are in the air is therefore twice the amount of time that you calculated for the way up:

$$2t = 0.4 \text{ s} \tag{19}$$

As a final example, add motion along the x-axis. Imagine that in addition to your 2.0 m/s in the positive y direction, you also had an initial velocity in the x direction. If you assume that your initial velocity in x, v_{xi}, is 0.5 m/s, how far in x will you travel during the course of this jump? Begin with a diagram, conveniently setting the point where $x_i = 0$ m to the location of the beginning of your jump.

48

Since you have not modified your y components at all, you can use the time calculated above for the full jump ($t = 0.4$ s) to apply to this problem as well. The full list of the x variables is therefore:

x_i: initial position along the x-axis = 0.0 m
v_{xi}: initial velocity in the x direction = 0.5 m/s
t: time between initial and final snapshots (for full jump) = 0.4 s
x_f: ?
v_{xf}: ?
a_x: ?

Do you have any more information? First of all, when you are jumping through the air, the only force acting on you is gravity, and gravity acts in the y direction. That means that you have no acceleration in the x direction. The exception to this would be if you were jumping into a stiff wind and had to take into account a force due to air resistance, but this is a force that you

49

can usually ignore inside the dance studio or performance space. This allows you to set a_x equal to zero. Given that you do not accelerate in the x direction, we know that you must maintain the same velocity in x throughout your jump. The final snapshot in time happens just as you are about to land, which means that v_{xi} and v_{xf} will be the same, both equaling 0.5 m/s. Our variable list therefore becomes:

$$x_i = 0.0 \text{ m}$$
$$v_{xi} = 0.5 \text{ m/s}$$
$$t = 0.4 \text{ s}$$
$$x_f = ?$$
$$v_{xf} = 0.5 \text{ m/s}$$
$$a_x = 0 \text{ m/s}^2$$

This latest list only contains one unknown, x_f, and it happens to be the variable you want to calculate. Consider once again the three equations of projectile motion to see which one will allow you to calculate x_f given the information that you have. The first equation is the only one that is useful. The second equation does not include x_f as a variable, and the third equation will result in your proving that $0 = 0$. (Try it!)

The first equation reads:

$$x_f = x_i + v_{xi}t + \frac{1}{2}a_x t^2$$

Plugging in the values that you know gives:

$$x_f = (0.0 \text{ m}) + (0.5 \text{ m/s})(0.4 \text{ s}) + \frac{1}{2}(0 \text{ m/s}^2)(0.4 \text{ s})^2 \tag{20}$$

The only non-0 term is the middle one, and you can see that your final x position equals

$$x_f = (0.5 \text{ m/s})(0.4 \text{ s}) = 0.2 \text{ m} \tag{21}$$

The amount of time that you are in the air is dictated by how high you jump, and how high you jump is dictated by your speed along the y-axis as you leave the ground. If you want to jump far, you need to balance jumping high with having some initial velocity in the x direction that will allow you to take advantage of the time that you are in the air.

Let's recap the procedure to follow when faced with projectile motion questions: First, make sure that you understand the physical scenario that you are working with by drawing a diagram and indicating the positions at which x and y equal 0. This sets your coordinate system for the problem. Also indicate which directions for x and y are positive. The convention is to set the positive x direction pointing to the right and the positive y direction pointing up.

Next, write down the full list of variables, filling in the numbers when they are known. Sometimes you will need to use information beyond what

is given directly in the problem. For example, remember that at the height of your jump, your velocity in y will equal 0. You also know that your acceleration in x is equal to 0 throughout your jump, which allows you to assume that your value of velocity in x will remain constant throughout the problem.

Then look carefully at your list of variables and note which unknown(s) you are trying to calculate in the problem you are solving. Armed with all of this information, look at your list of equations to see which contains your unknown variable. If there is more than one option, pick the one that appears to have the simplest calculation associated with it. As you do the calculation, keep an eye on your units, making sure that at all times the units to the left and right of your equals sign match.

These equations of motion only work under conditions where acceleration is constant, so they cannot apply to a full dance, during which the dancer interacts with frictional forces in complicated ways. But they are useful tools with which to analyze the motion of dancers in the air.

Zero Velocity

We have considered jumping straight up and down and skipping up and outward in this chapter, but of course the spectrum of ways that dancers jump is much broader and can require more specialized training. The means of take-off and the position in the air of the legs, arms, and torso have great variability. A dancer known as a "jumper" has the virtuosic ability to soar through the air much higher and farther than the average person. Choreographers find ways to take advantage of this spectacle. And when they do, what they are actually capitalizing on is the moment in a jump that physics defines as "zero velocity": when the object, or in this case the dancer's body, that is moving upward against the pull of gravity slows to a stop at the height of a jump, before falling back to the earth.

In David Parsons's 1982 piece *Caught* a strobe light illuminates the apex of a dancer's jumps. The dancer's preparation, take off, and landing are concealed by the otherwise darkened stage; the audience sees only a series of climactic images of the dancer in flight. Or, in the finale of Balanchine's 1947 ballet *Symphony in C*, fifty-two dancers perform a *petite allegro*, a quick series of small jumps, which carries them all into the air at once. The entire stage appears to be ready for liftoff.

Physics tells us more about what is actually happening at the height of those jumps. What the science cannot tell us is what those moments *mean*. To those watching, the dancers appear, however fleetingly, to have escaped the laws of nature.

4. Friction

Imagine that you are standing at one end of a large ballroom with polished wood flooring. If you want to get to the other end as quickly as possible, would you have an easier time starting to sprint in socks or wearing your sneakers? Clearly, sneakers would be more effective because they would provide more friction. While friction is often cast as the force that slows us down, it is just as central to starting our motion as it is to stopping it: you will need friction in order to get going.

Friction acts upon dancers' bodies at every turn, whether imposed by the floor, shoes, sets, or costumes, or even by the air. Dancers require friction to execute nearly every movement, including jumps. Just as the best dancers and choreographers develop distinctive relationships with gravity, so must they contend with friction.

Dance employs friction for expressive power. As tango dancers sweep their legs in a circle, for instance, they sense the friction between their feet and the floor, and intuitively apply a degree of pressure to create the precisely cut shape. Tango has a tensile, elastic quality created partly by the flicking, sweeping action of the feet in conversation with the floor. None of these qualities would be possible if material surfaces slid by each other with an unchanging degree of resistance. In dance, friction produces meaning.

Let's get moving.

Improvising with Friction

With so many natural forces acting upon our bodies at once, it is impossible to focus exclusively on friction. But isolating friction can give us a motif to use in our movement practice: we can adjust our awareness and manipulation of frictional forces as we move.

Find an open space—a dance studio or a gym—with a smooth floor. Work in either socks or bare feet, to start. You might opt to change your footwear midway through the exercise. You may wish to use music. (Jelly Roll Morton's solo piano recordings are among our favorites for this investigation. His textured, swinging downbeats exemplify the varied dynamics you are about to explore.)

Start by focusing on the encounter between your feet and the floor surface. Keep in mind the various ways your feet can move that were explored in the previous chapter: heel first or toe first, with a relaxed or pointed foot, rolling through or flat-footed—try out all the options. This exercise can also be done with the hands or while you are lying on the floor, depending upon your physical needs.

Begin to discover ways to slide and skate, press and dig into the floor. Shift your weight from one foot, or one hand, to the other as you work. Take stock of your upper body: relax your neck, arms, and shoulders and release your weight downward—recalling what you have learned about gravity.

Try out a few approaches: alleviate the pressure, and you can glide over the floor surface. Drive the pressure downward, on the other hand, and your movement is impeded. Skate around the space, varying the timing. Don't worry about trying to adopt specific positions, or what your work might look like to an observer. We used time to structure your movement studies in previous chapters; in this case, your attention to and engagement with frictional forces structures this exercise.

Because this is the first movement study that works with improvisation, we will set one more helpful limit: set the timer for eight minutes (our by-now familiar eight minutes). One more detail: remember to move backward as well as forward … and go.

The movement research in this exercise consists of sharpening your attention to the sensation of friction in relation to the force you apply into the ground, and using this attention to inform your movement choices. Listen to your body: the stickiness or slickness of the floor affects your physical organization, coordination, and movement quality. This is the reason why dancers can be very picky about their footwear and the quality of the floor surface, just as tennis pros are picky about their sneakers and court surfaces.

"Sharpening your attention" in this context means building sensorial knowledge and muscle memory. Your body will remember the purposeful interaction with friction. That relationship is important, for it will ultimately help you to perform different movements with different qualities: smooth and edgeless, or quicksilver and sharp.

Kinetic Friction

While you were improvising, you were directly engaged with the phenomenon of friction. You must contend with friction daily, whether or not you are dancing. The kinesthetic intelligence that you are beginning to hone taps into the two categories that physicists use to classify friction:

Kinetic friction: The force due to friction when two surfaces are in motion with respect to each other.

Static friction: The force due to friction when two surfaces are at rest with respect to each other.

It usually requires less force to *keep* an object moving than it does to *start* it moving. That makes these two categories of friction necessary: for two materials in contact, the kinetic friction between them when they are moving is less than the corresponding maximum static friction.

Let's build the mathematical model—in this case a formula—for kinetic friction. The formula should include the variables that have an impact on the force. What determines how strong the force of kinetic friction will be?

You know from your recent movement research that the nature of the two materials in contact affects the force of friction that is present when they move with respect to each other. *Socks & floor* will have very different friction from *skin & floor*, which is different from *socks & wet floor*. Information about the material combination must therefore be included in the formula. It is possible to get a good sense of the force strength by including a single number associated with each pair of materials, named the coefficient of kinetic friction for the pair. This one number encodes all of the information you need about the nature of those two specific materials in contact. Physicists use the symbol μ_k to represent this value.

The coefficient μ_k will be a number between 0 and 1. The number is close to 0 for pairs of materials like teflon and a rubber spatula, for which we can expect the surfaces to glide easily over each other. The number is close to 1 when the materials do not easily glide over each other—material combinations such as iron and steel or rubber sneakers on a gym floor.

But more information is needed to complete the mathematical model of kinetic friction.

You can break down the movement exercise further with an eye toward the physics. Standing in socks on a slippery surface such as smooth tile or polished wood, put most of your weight on one foot and slide your other foot around on the floor, just brushing the surface. Your foot should slide easily. If you start to shift more and more of your weight onto the moving foot, what do you experience? As you increase the magnitude of the force between your foot and the floor you should feel more resistance to the foot's motion. Increasing the force pressing the two materials together increases the force of friction.

To account for this, your mathematical model will need to include the force between the two objects in contact, not just the types of material that are in contact. For clarity, label the two materials in contact as A and B. We know from Newton's 3rd Law of Motion that the magnitude of the force that surface A exerts on surface B is equal to the magnitude of the force that surface B exerts on surface A. The perpendicular component of the force exerted on one surface by another can be expressed by the *normal* force, in which *normal* does not mean ordinary but perpendicular. The strength of that force, which dictates how strongly the materials are pressed into each other, is related to the magnitude of the force of friction. Physicists use the variable F_N to represent the normal force.

The complete expression for the model of the force due to kinetic friction, taking into account both the nature of the materials in contact through μ_k and the force between them, is

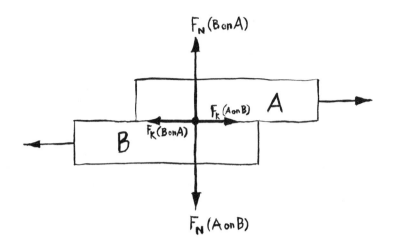

$$F_k = F_N \mu_k \tag{22}$$

where F_k is the variable for kinetic friction. We have not yet discussed the direction of the force. If you try to slide your foot forward, the force from friction pushes backward, against the motion. If you try to slide your foot in the opposite direction, backward, the friction force will point forward, again opposite to the direction of motion. The force due to kinetic friction always acts in the direction that opposes motion.

Friction, on the microscopic level, is extremely complicated. It depends on the roughness of the materials, the speed at which they move with respect to each other, and the nature of the chemical bonds that are formed between their molecules.

In this macroscopic model, which works well for dancers moving through the world, you did not have to include in the equation the amount of surface area in contact between the materials, or how quickly they are moving with respect to each other. Only two parameters are needed to calculate the value of kinetic friction.

Surfing

In this next movement exercise, you can begin to explore the ways that dancers cultivate an intimate relationship to friction. Often, this relationship revolves around the interaction between the footwear and the floor. Dancing barefoot requires a different degree of pressure into the floor from dancing in socks or sneakers. Specialized footwear—from pointe shoes to tap shoes—requires specialized floor surfaces, with different coefficients of friction. Depending on the dance form, different degrees of resistance are optimal: some dancing relies on an easy slide, other dancing makes use of greater resistance. And as you have just learned, both the amount of force

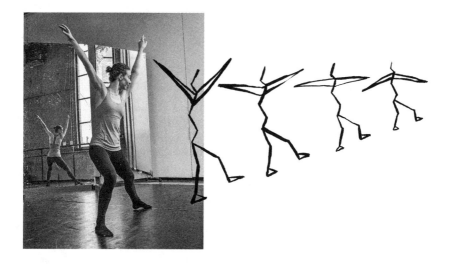

pushing the surfaces together and the coefficient of friction will affect the dancer's sensorial experience.

In the next movement study, we will ask you to manipulate coefficients of friction through footwear. To do so, we will take a cue from the seminal dance *Deuce Coupe* by the choreographer Twyla Tharp. Commissioned by the Joffrey Ballet to create a new work, Tharp mixed her eclectic modern dance style with classical ballet and set the entire extravaganza to music by the Beach Boys. With exacting precision, a ballerina performs a series of ballet steps while a horde of modern dancers shakes and shimmies around her. When *Deuce Coupe* premiered in 1973, no such crossover ballet had ever been seen before.

In one memorable section, the dancers take running starts and "surf" across the stage as if riding waves ashore, notably accompanied by the Beach Boys hit "Catch a Wave." They can't travel terribly far, due to—you guessed it—the restrictive levels of kinetic friction, which ultimately stops their motion. But they do slide far enough to create the convincing illusion of surfing, in an exemplary choreographic exploration of frictional forces.

In this exercise, try out the idea of surfing. You will need an open space, preferably a dance studio, though a gym floor or ballroom would suffice. The basic step involves running and sliding across the floor—you may have tried this as a child.

Play the Beach Boys song "Catch a Wave" as you work (this is key!) Try the movement first in socks, then barefoot, then in rubber-soled sneakers.

Note the details: Where is your center of mass in each version? Does the effect change if you slide on one foot or two? In what ways do the different kinds of footwear impact the duration and length of travel in a slide? Not surprisingly, you will find that it is much harder to slide in sneakers than in socks.

Here's a surprising detail that you will discover from the physics: the frictional force you experience while surfing does not depend on the total area of contact between your body and the floor. Although it might seem as though surfing on one foot should slow you down exponentially less than surfing on two feet, in fact the total resistance due to friction that you will encounter remains the same.

To understand why, it helps to remember that in the formula for kinetic friction, you only need to know the coefficient of kinetic friction, determined by the type of materials that are in contact and the normal force between your feet and the ground. If all of your momentum is horizontal, the normal force between your feet and the ground will be determined by your weight, and your weight does not depend on how many feet you have on the ground while you surf.

Static Friction

Grinding to a halt while "surfing" across a floor is a process in which kinetic friction acts to slow the body's velocity. It is a deceleration, or negative acceleration, because the acceleration is in the opposite direction of your motion. Once you have slid to a stop (standing upright, bent forward, toppled onto the floor, or however you may find yourself), friction no longer operates; it does not come into effect again until the moment you want to get going once more. At that point, you have to contend with static friction.

When you begin to walk, run, or dance across the floor, it is only thanks to friction that you are able to achieve the net force that pushes you forward. When you try to start running in your socks, the amount of push forward you get is reduced if your feet slip, but not if they remain firmly planted. What governs how much you can push before slipping? What gives you traction? This force is called static friction.

The magnitude of the force of static friction, like kinetic friction, depends on the nature of the two materials that are in contact, as well as on the normal force between them. The following equation dictates the maximum force that static friction can provide ($F_{s,max}$) before slipping occurs and kinetic friction takes over:

$$F_{s,max} = F_N \mu_s \tag{23}$$

If you stand still on a Marley floor—the non-slip vinyl material often used in dance studios—wearing rubber-soled shoes, the balanced forces of gravity and the floor pushing up on you keep you at rest. As you begin to push off of the ground with one foot, the floor can respond with an equal and opposite force. The force you apply will be at a diagonal with respect to the ground. That vector can be broken up into a component parallel to the floor and a component perpendicular to the floor:

The component that is pushing directly into the floor is the normal force, because it is perpendicular to the surface. The normal force F_N is what is needed in the equations to calculate frictional forces. The force that is parallel is opposed by the force due to static friction that pushes back on you. As long as your parallel force is equal to or less than $F_N \mu_s$, your foot will not slip.

For a given pair of materials, the coefficient of static friction μ_s tends to be higher than the coefficient of kinetic friction μ_k, consistent with our experience that it is more difficult to get an object moving than to keep it moving. Put another way, in most cases, $\mu_s > \mu_k$. We list here a few examples of coefficients of friction for pairs of material:[9]

Material 1	Material 2	μ_s	μ_k
Rubber	Concrete	1.0	0.80
Glass	Glass	0.94	0.40
Steel	Steel	0.74	0.57
Waxed Wood	Wet Snow	0.14	0.10
Teflon	Teflon	0.04	0.04
Joints (synovial) in people		0.01	0.003

This table shows the coefficients of static and kinetic friction for different pairs of materials and for the synovial joints of people with healthy cartilage. Note that the coefficients are closer to 1 where there are higher frictional forces and closer to 0 where friction is low.

Even if you are moving around in the studio, as long as your shoes (or socks or feet) are not slipping with respect to the floor when they are in contact, your motion is enabled by static friction. The same can be said for motion with wheels. If you are in a wheelchair or automobile, the part of the wheel that is in contact with the ground at any given time is not slipping, so static friction is at work. Kinetic friction is the force engaged when two surfaces are sliding with respect to each other. And when wheels slip—maybe because you are stuck in mud, where the frictional forces are low—it becomes difficult to move.

Calculating Frictional Forces

The formulas for frictional force calculations need only a few variables. These include the normal force F_N that pushes the two materials together, and either μ_s (if the two materials are not slipping with respect to each other) or μ_k. Your main task will be to get from the physical scenario that you are trying to model to the point where you can use the formulas.

In order to do a variety of calculations, you first must become comfortable with breaking a force up into its parallel and perpendicular components. If you are jumping straight up, the force you apply to the ground will be only in the vertical direction. It will equal the normal force F_N between your shoes and the floor. But if you are jumping out or starting to walk or run, you will push down at an angle, with some component *perpendicular* to the floor—contributing to the normal force F_N—and some component *parallel* to the floor. The parallel component F_P is what will result in the floor propelling you forward.

How would you take a force acting at some angle and break it into components? Let's use the diagram on the left to construct the triangle on the right:

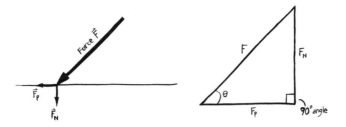

This is a "right triangle" because two of the edges join to form a 90-degree angle. We have labeled the angle that your force makes with the floor θ. Your next job is to label the sides of the triangle itself. It is time to bring trigonometry to dance!

The edge across from the 90-degree angle is the *hypotenuse*. The side that joins θ and the right angle is the *adjacent* side because it is next to the angle the force makes with the ground. The remaining side is the *opposite* side because it is opposite to the angle the force makes with the ground.

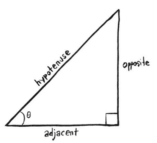

With your triangle successfully labeled, you will need your calculator and this handy trigonometry chart, which defines sine, cosine, and tangent with respect to the labeled triangle:

$$\sin \theta = \text{opposite/hypotenuse}$$
$$\cos \theta = \text{adjacent/hypotenuse}$$
$$\tan \theta = \text{opposite/adjacent}$$

Let's give this a try. Imagine that you are in your rubber-soled sneakers standing on a concrete floor. You push into the ground at an angle of 40 degrees and with a force of 600 newtons (N). Will your feet slide? Or will static friction hold?

First, draw a sketch of the situation with the information you have been given:

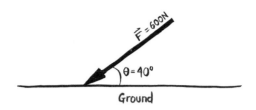

Force you apply to the ground

$\vec{F} = 600N$

$\theta = 40°$

Ground

Then, make a force triangle based on the situation. You will find that your 600 N force is the hypotenuse, your adjacent side is aligned with the ground and gives you F_P, and the opposite side is aligned with the normal force F_N. Using trigonometry, you can see that the sine of the angle will equal the ratio of the opposite side to the hypotenuse:

$$\sin 40° = \frac{opposite}{hypotenuse} = \frac{F_N}{600 \text{ N}} \tag{24}$$

You can simplify the equation to the following after calculating the value of $\sin 40°$ to be approximately equal to 0.64:

$$0.64 = \frac{F_N}{600 \text{ N}} \tag{25}$$

Solve for the normal force by multiplying both sides of the equation by 600 N:

$$(600 \text{ N})(0.64) = F_N = 384 \text{ N} \tag{26}$$

Given the normal force and the coefficient of static friction for these two materials, what is the maximum parallel force that static friction can respond to without slipping? The coefficient of static friction μ_s for rubber and concrete is 1.0. You can plug that into your equation for the maximum force from static friction along with your calculated normal force F_N of 384 N:

$$F_{s,max} = F_N \mu_s = (384 \text{ N})(1.0) = 384 \text{ N} \tag{27}$$

If the coefficient of static friction had been a smaller number, it would have given a smaller maximum value of the force due to static friction for the normal force. So, will you slip? What parallel force F_P are you applying to the concrete?

$$\cos 40° = \frac{adjacent}{hypotenuse} = \frac{F_P}{600 \text{ N}} \tag{28}$$

Plugging $\cos 40°$ into your calculator, you will find that it equals approximately 0.77. You can therefore simplify the equation like this:

$$0.77 = \frac{F_P}{600 \text{ N}} \qquad (29)$$

Solve for the parallel force F_P by multiplying both sides of the equation by 600 N, to find:

$$(600 \text{ N})(0.77) = F_P = 462 \text{ N} \qquad (30)$$

Uh oh. $F_P > F_{s,max}$! The parallel force is bigger than the force the sneaker/concrete pair can withstand without slipping. It turns out that you need to be at an angle of at least 45 degrees with the ground in this scenario in order to not slip. The angle required in order to not slip will be different for other pairs of materials.

If People Were Boxes

Physicists often teach friction by presenting the image of boxes sliding down inclined planes. In honor of this convention, we propose a movement study that replaces boxes with people. Consider what follows an example of the choreographic imagination of a physicist at play in a dance studio ...

Pair up. One person will lie on the floor and the other will remain standing. If you are the person on the floor, assume the starting position that was used in an early gravity exercise: lying flat on your back, legs extended forward, slightly wider than a hip's width apart, arms flat several centimeters away from the sides of your body. Take a moment to relax each muscle: you may feel your body begin to melt into the floor.

If you are the person standing, face the feet of the person on the floor. Bending down, gently place your hands under the ankles and lift his or her legs about a meter off the floor. Be sure to lift using your own legs as a firm base of support. Relax your arms enough to feel the weight of the person's legs in your hands. Gently pull away from his or her body. The person on the floor should in turn relax, and not grip or hold on to unnecessary tension. If you are on the floor, release the weight of your legs and relinquish control to the other. If you are standing, experience the weight of your partner's legs.

Now explore together the threshold of force required to overcome static friction. The person standing is in charge. Begin by pulling gently. Pull just enough to feel the resistance of your partner's body to motion. Then, while anchoring your legs and maintaining a solid open stance, pull with a greater force that allows you to slide your partner gently along the floor. Just a few centimeters will do—enough to allow you both to experience the sensation of overcoming static friction and entering into an engagement with kinetic friction.

Dancers do not usually drag each other around like this, but the exercise gives you a sense of how frictional forces feel and act upon your body. The difference between people and boxes is that people can experience a sensorial engagement with frictional forces that boxes cannot.

Composing with Friction

Paying attention to the sensation of frictional forces can become a tool not only for dance techniques but also for choreographic composition. The finest movers are innate choreographers: they rely on their ability to transform static into kinetic friction and take full advantage of the hiccups of static friction that can occur in the midst of motion. Take, for instance, the pratfalls of the silent film star Buster Keaton. His characteristic slips, trips, tumbles, and falls off of buildings, furniture, cars, carts, or dusty hills read like an epic paean to the force of friction. (You can find clips of his physical virtuosity online.)

The presence or absence of frictional forces can change the qualities of a dance. Watching the flamenco dancer Israel Galván, a viewer can feel the visceral impact of his dancing. Sequencing heel-toe, toe-heel, Galván unleashes quick-fire footwork with panache. One rhythmic fusillade folds into the next. He strikes the floor, dramatically stopping a phrase before continuing into another movement idea. What he is actually doing is playing with frictional forces. When he drives his foot diagonally downward into the floor and momentarily halts, kinetic friction becomes static friction. The force punctuates his dance: his expressiveness relies on friction.

The absence of static friction is another compositional tool. A movement phrase could be lifted entirely off the floor through ropes and rigging, thereby negating the effect of friction on the body. The choreographer Trisha Brown began her Monteverdi opera *L'Orfeo* with a remarkable flying solo: the dancer swims, dips, and hovers, deploying Brown's aqueous movement style in the air. With nothing to push off of and no resistance aside from negligible air resistance, the dancer finds new resources with which to move. Place the same phrase back on the ground and it would look very different.

Choreographers sometimes alter the coefficient of friction by changing the texture of the floor on which the dancers perform. One such choreographer was Pina Bausch, a pioneer of a form referred to as *Tanztheater*, a genre known for combining theatrical and dance elements. With her set designer, Peter Pabst, Bausch created dances for stages covered with water, soil, flowers—materials that undeniably establish atmosphere but also present a range of frictional encounters. Dancers will move in a pool of water differently, for instance, from the way they move in a mound of soil, because the substance requires dancers to modify their attack and execution. The same movement phrase will look different if performed on a slip-and-slide than in a pool of molasses.

Playing with the ways that substances affect the dancers' imaginations and kinesthetic response is means of changing the *quality* of the movement. But not every choreographer needs a mound of dirt to alter the way people move onstage. We are going to try out another choreographic tool that uses *level changes* to engage with friction.

This exercise involves changing the orientation of movement in space. You will need a movement phrase—a short sequence of movements—to

carry out this research. The simplest one we have done at this point is four plies in a row: two counts to bend your knees, two counts to return to standing. You designate a "front" simply by choosing a direction in which to face. The movement is accomplished upright, with your spine perpendicular to the floor. (If you have other more complex movement phrases available to you, use them.)

Now, change the orientation of the phrase by lying on the floor and choosing a new "front": either the ceiling, the floor, or one side of the room. Depending on your starting position, you may be lying on your back staring up at the ceiling or facedown into the floor, or lying on one side or the other. Work through the phrase in this new orientation, trying to remain as faithful as possible to the original phrase within the unfamiliar conditions.

You will need to adjust the phrase to accommodate the new situation: you may discover that your points of friction and external forces change, and that your arms and legs might need to work in new ways in order to execute the material while lying on the floor. A degree of choice inevitably enters in when you begin to experiment with your material, even when you diligently attempt to preserve the integrity of the phrase.

Changing your orientation alters your relationship to gravity as well as to the frictional forces that act upon your body while you perform the phrase. When you are in greater contact with the floor's surface area, each action and gesture should feel different—not least because the floor offers greater resistance than the air. Repeat the phrase at least five times to fully explore this new texture in the movement. You are researching the quality and sensation of friction, and its impact on your physical expression.

This movement study also draws you into the realm of *choreographic re-*

search. In tinkering with fundamental aspects of a movement phrase, a chore-ographer looks for compelling new movements and textures to incorporate into a final composition. The movement phrase serves as some choreographers' raw material, just as a mound of clay might be the starting point for some sculptors. Later in the book, we will look at other elements—namely, energy, space, and time—that a choreographer can treat and transform in the process of researching a movement phrase.

Level changes are one way a choreographer creates variation in a movement phrase. Levels range along a dancer's vertical axis, from lying down to rising onto the toes to jumping. Dancers raise and lower themselves in elaborate ways: by falling or dropping to the floor or popping up into a jump from a deep plié. Such strategies defy easy explanation (and in some cases must be experienced to be believed!). But no matter how complicated the choreography becomes, static and kinetic friction are engaged at every moment, from laying a movement phrase on the floor to sliding dancers across a stage.

Tango in the World

Imagine dancing in two different worlds in which you sense opposite extremes of frictional force. In a world in which you sense very little friction, a partnered tango would feel slippery—it would be impossible, for you could only approximate the position, with no ability to lock into place with your partner. Then imagine a world in which you sense such a high degree of friction that sharp, quick motions would be easy. The ground would grip your feet, launching you in any direction and keeping you from slipping. But when you made contact with another body in order to tango, your skin would peel off from simply brushing against the other person.

The strength of the human body is physically calibrated to the earth's surroundings—the air, the ground, the material surfaces with which we interact daily. Human psychology would be very different if the conditions of friction dramatically altered. Our intimate relationships would change if we slid past each other with every attempt to make physical contact: consider the slippery tango or a slip-and-slide waltz. The fact that we can grip and hold on to another body with reasonable effort without lashing ourselves together or having to pry ourselves apart makes certain dance forms possible. The frictional forces that enable dance dictate how we relate not only to our environment, but also to each other.

5. Momentum

What does it mean to be a good dancer? What is good art? What is good science, for that matter? How can you be both a good dancer and a good scientist at once?

These questions lead us to *virtuosity:* a tricky and contested concept that invites us to consider what is required to perform our disciplines well. In dance, many define virtuosity as technically impressive physical feats, such as the ability to jump higher and spin faster and longer than the average person. A more nuanced definition of virtuosity in dance, however, will take into account a performer's intensified attention to building momentum through movements big and small.

A great dancer such as Mikhail Baryshnikov, a Russian ballet star of the late twentieth century, develops a sophisticated relationship to natural forces. In his youth, Baryshnikov marshaled momentum to shoot across the stage in a series of astonishing leg beats, or to transform a complicated pirouette into dips and spirals. Having performed now for over fifty years, he has converted the brasher energy of his ballet days into potent theatrical gestures. Baryshnikov's momentum is a physical concept with metaphorical resonances: he not only understands gradations of muscular force; he also understands live performance, in which the performer exerts an energetic grip on the space that carries the performance along. His virtuosity lies in the intense awareness with which he transforms one movement or state into the next.

Physicists define momentum as a measure of an object's mass times its velocity. Not unlike Baryshnikov, skillful physicists are able to move this concept and formula across circumstances—with an eye on which details are essential and which assumptions can be discarded to analyze the new conditions. Physicists apply momentum not only to motions that people can perceive in their everyday lives but also to subatomic activity. In particle physics, for example, tracking conservation of momentum is especially useful for examining high-energy collisions, in which particles collide and fly apart. Knowing that momentum is conserved helps physicists to analyze the decay.

Momentum influences everyone, but few people know how to wield the physical effects or conceptual implications of momentum as virtuosically as dancers and physicists.

Catching a Wave

Dancers use momentum to fold one movement into another. This momentum works in stages: first, there is a marshaling of forces, which must come from somewhere—often initiated in a particular part of the body or built into a sequence of choreography. Some dancers think of this moment as the "bottom" of a jump or an action. This initial movement then generates a force that the dancer rides into the next action. Within the fulfillment of each movement lies the potential for the next movement that occurs. Imagine an ongoing series of waves, with crests of energy and dips of recovery that turn one wave into the next.

Momentum is especially important when transitioning between movements. Generating and conserving force helps dancers to connect a series of jumps, for example, or link together movements that flow from one level to another. Even a simple set of gestures requires thoughtful force, whether modulated or more aggressive, to carry the dancer through the sequence.

Different dance forms treat momentum differently. One technique might require a jump to come out of nowhere: try, with very little lead-in, to launch yourself up off the floor. Another might oppose the effects of momentum, by requiring a dancer to maintain a more rigid torso, for instance, while turning. In contrast, many of the contemporary choreographic styles developed since the 1980s focus on a liquid sequencing and flow within the body. Rather than opposing the direction of force, the dancer catches the wave and allows the momentum to generate the next movement that comes. The best dance improvisers work like jazz musicians, following and altering lines of natural force like riffs on a melody.

The force of gravity upon the human body provides dancers with one of their greatest sources of momentum.

By way of inquiry, try this movement exercise. Swing your arms along a 360-degree circle parallel to the floor. This simple exercise highlights the sensation of momentum on the arms.

Wind up by pulling the left arm behind the body to the left, which draws the right arm across the body and gently twists the upper body at the waist to the left. Following this windup, lead with the right arm back around toward the right, in a swinging motion. The arms and torso travel in this repeating pattern for the rest of the exercise: the arms swinging to the right, then the left, then the right, and so on. Add a slight plié in the legs at the moments when your arms are wrapped around your body and straighten them when your arms swing back the other way—you'll find that the legs help lend additional momentum to the swinging of the arms. Start out slowly and develop a steady rhythm. When you begin to feel comfortable with the swinging sensation, accelerate incrementally. You may discover that the force you feel in the plié and the swinging of the arms will intensify as you increase your speed.

Reading the description above will not fully convey to you the sensation of momentum if you have never applied your attention to experiencing it.

So now put down this book and give it a try. Slowly increase the speed of your swinging until you cannot possibly swing faster, and then incrementally slow back down to a rest.

You may find your fingertips turning slightly red and feel a tingly sensation as you work. In reaction to the acceleration, the blood rushes outward to your extremities. You may also have noticed that you have the option of a good deal of variation and play with the forces. You can change the depth of the plié and the engagement of the legs or the height of the swing. One variation is to allow the arms to swing in an elevated arc as they pass from one side of the body to the other. This imparts a sense of loft—you may feel as if your arms are a pendulum that rises and dips, alternating from one side to the other.

The windup of the arms, the slight plié of the legs, and gravity's pull all fuel the swing. Your swinging arms could pull you around in a turn, switch your direction, draw you to the floor—many different outcomes are possible.

Efficiency in dance results from the dancer's knowing precisely where and how to generate momentum and how to engage with the forces that result. Consider the legs, spine, head, and torso, in addition to the arms: all of these hold the potential to provide a kind of momentum that the body must then respond to and utilize. We could spend time analyzing this activity, as many dance teachers and choreographers do, but it is foremost a sensation. Dancers get to know the concept of momentum through motion.

Calculating Momentum

Momentum is a central idea in physics, just as it is in dance. The two disciplines use the term in similar ways. For example, the more momentum there is in a movement, the more difficult that movement is to stop. We assign a numerical value to momentum in physics, and this enables a surprising number of useful calculations that teach us about our potential for (and the constraints upon) movement. Momentum is also conserved under certain circumstances. By *conservation*, we mean that the quantity remains constant from one moment to the next, and this applies a powerful constraint on motion.

The momentum of any object is equal to the object's mass multiplied by its velocity. Physicists use the letter p as the variable to denote momentum.

$$p = mv \tag{31}$$

Momentum is a vector quantity: it has both a magnitude and a direction. The direction of momentum is determined by the direction of the velocity of the object. The magnitude of momentum is determined both by the magnitude of the velocity—also known as the object's speed—and the amount of mass that an object has.

It can be useful to think about the maximum and minimum values of momentum that you can attain under your own strength. Given that you exist and are reading this, it is safe to assume that your mass is not equal to 0. This means that the only way for your momentum to be equal to 0 is if you have zero velocity. And what about your velocity? You are hurtling around the sun with a substantial velocity. The sun, and our entire solar system with it, is moving through our galaxy, and our galaxy is moving with respect to other galaxies in the universe. Which velocity should we use?

There is no such thing as an absolute velocity, so we need to define velocity with respect to some frame of reference. Since we are primarily interested in the movement of bodies in the context of dance, we will consider velocities with respect to the dance studio or the room in which the bodies are moving, unless otherwise noted. In this reference frame, the only way to have zero momentum is not to move with respect to the floor. (The mass does not depend on the reference frame.)

We have seen how you can minimize your momentum: simply do not move. How do you maximize your momentum? Looking back at the equation, we can see that there are two inputs to the calculation of your momentum: your mass and your velocity. Your mass is constant from one moment to the next, so your best option for maximizing momentum in the short term is to increase your velocity. For an extreme example of a human with self-generated high momentum, let's look at the Jamaican athlete Usain Bolt's world-record-breaking 100 m race at the 2009 World Championships in Berlin. He covered the distance in 9.58 seconds, which results in an average velocity of:

$$v = \frac{distance}{time} = \frac{100 \text{ m}}{9.58 \text{ s}} = 10.4 \text{ m/s} \tag{32}$$

Given his mass of approximately 95 kg and velocity of 10.4 m/s, his average momentum during the race was approximately:

$$p = mv = (95 \text{ kg})(10.4 \text{ m/s}) = 988 \text{ kg m/s} \tag{33}$$

Perhaps you are not quite that fast. If a person is able to achieve a maximum velocity of 8 m/s, what mass would he or she need to have in order to reach the same magnitude of momentum? Since we want the same value for p, but v has decreased, we would expect the mass m to increase to make up for this.

We can rearrange this equation

$$p = mv \tag{34}$$

to

$$\frac{p}{v} = m \tag{35}$$

Substituting the numbers for the variables, we get:

$$\frac{988 \text{ kg m/s}}{8.0 \text{ m/s}} = 123.5 \text{ kg} \tag{36}$$

In order to denote a direction for your momentum, you will need to define a coordinate system with an x-, y-, and z-axis and define the directions in which the axes are positive.

What does it take to maintain a non-zero momentum for an extended period of time? How long could you maintain a speed of 1 m/s inside a room? At some point you would need to stop and rest. We have to work non-stop to maintain a non-zero velocity.

To explore this idea, try this movement exercise while maintaining a focus on the laws of physics. Try jumping forward, using all of the dance knowledge you gained in Chapter 3. You leave the ground with some momentum thanks to the strength of your legs, the force you have applied to the floor, and the floor's equal and opposite force back on you that propels you up and forward. As soon as you leave the ground, you lose the force of the floor on you that counteracts the force of gravity. Your upward velocity will immediately begin to decrease. If you could turn that gravitational force off, your momentum would be maintained, and you would float away from the earth's surface.

Gravity cannot be turned off. However, it only acts to pull you toward the center of the earth—it doesn't change your forward momentum. Unless you jump into a strong wind, the component of your momentum parallel to the flat ground will remain constant, with the exception of air resistance, which is often negligible. When you land, the floor abruptly applies a force that opposes your forward momentum, stopping you whether gracefully or not, depending on how you make contact.

Conservation of Momentum

Momentum is a conserved quantity within a system if there are no external forces acting on the system. The momentum at any specific point in time is equal to the momentum at all other times in a system that is isolated from outside forces. Snapshots taken of a system at any two moments of time, labeled with the letters i for initial and f for final, would yield this equation:

$$p_i = p_f \tag{37}$$

That is, the initial momentum equals the final momentum for this system without external forces. Substituting the definition of momentum from equation 34 for p gives:

$$m_i v_i = m_f v_f \tag{38}$$

Physicists use this formula for the conservation of momentum to study everything from subatomic particles to a leaping dancer to a system of binary stars.

Let's return to your attempt to maintain a constant speed—and therefore constant magnitude of momentum—in a room. With each step or jump you take, forces from gravity, the floor, and even the air are acting on you.

Humans have very little experience with a true zero-external-force environment. Hypothesizing such conditions, however, can be a useful guide for thinking through the implications of conservation of momentum within physics. To get you into a situation in which you do not have external forces acting upon you, we will need to put you deep into outer space. But first, we will return to our movement research on earth.

A Momentum Movement Phrase

We feel momentum acutely when performing extreme physical actions, such as jumping. But we also experience the sensation described by $p = mv$ in everyday activities. Change your direction while walking, and you have engaged with momentum. Try running and then changing your direction, and the forces you experience will feel even greater. (By now, you know that friction and Newton's laws are at work, allowing you to redirect your motion.) We can use these ideas to build a movement phrase based on the concept of momentum.

Start with the legs: step forward with your right foot, then your left, then your right. So far, so good, right? Just like walking. Now it gets trickier: increase your speed so that you are running along what we will designate as the positive x direction. Next, you are going to pivot a half-turn to the right, ending up on both feet, leaning forward at an angle to face the direction you just came from. Try to rise onto the balls of your feet as you pivot, and look for the feeling of falling forward, with the upper body angling more steeply than the legs. From here, repeat the sequence: run forward again, this time in the negative x direction, then, quickly pivot a half-turn to the left to return to face your original positive x direction. Then start the phrase all over again: run, pivot, run, pivot, run, pivot ...

When you move through this "momentum movement" phrase, try to hover in the pivots, using your arms to help you rise. You are looking for moments that feel almost weightless, a momentary and fleeting suspension that comes about not through pulling yourself up, but through turning the momentum of your run into another action. You can also explore the force of momentum that swinging your arms adds by giving you additional pressure into the floor.

Note that when you lean your center of mass is entirely shifted forward, outside of the bounds of your area of support. Your head and your feet are not aligned. You cannot remain in this position for long—you are actually falling. But the fall gives you the momentum to continue the phrase.

Dancers performing a movement phrase can feel as though they are traveling through a topology of energies, and each individual body identifies, sorts, and makes decisions about the most effective means of marshaling momentum, toward the most fully actualized execution. No two dancers necessarily locate their resources in exactly the same place.

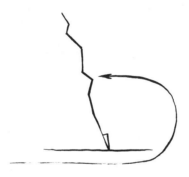

Conservation of Momentum in a System

Once you have mastered the momentum phrase above, it is time for us to throw another variable at you: outer space. You are stuck in outer space 10 m from your spaceship. You are trying to perform the momentum phrase in order to make your way back to the ship, to no avail. You tread in place, fling your arms, and deploy everything that you learned by practicing that phrase in your earthbound dance studio, but nothing gets you closer to your ship.

Then you remember Newton's 2nd Law of Motion: you need a net external force on your body in order to accelerate in the direction of the ship. You look around frantically for a surface to push off against but see only the flashlight that you are holding to light your way in deep space. Help is in your hand.

In order to understand how the flashlight can save you, you need to understand the idea of conservation of momentum, which holds as long as no external forces are acting on a system.

Let us back up and examine your circumstances more carefully. Are there any forces on you as you float near but not within reach of your spaceship, holding your flashlight? You will experience an external force due to the gravitational attraction between yourself and the ship, and the ship will experience an equal attractive force pulling it toward you. We will assume that the ship is not massive enough to pull you together before you run out of oxygen. We will therefore consider this situation to be an approximate zero-external-force environment.

How can you get your body back to the safety of your spaceship? If you were on the earth you would take advantage of contact with the ground between you and the ship, and the friction between the ground and your shoes, to propel you toward the ship by pushing your feet into the ground and allowing the ground to push you forward—a motion otherwise known as walking. But in deep space, where there is no planet to make contact with, you cannot rely on friction with the ground. It is just you, the flashlight, and conservation of momentum.

The momentum of the system of you and the flashlight is initially 0, because each of you has zero velocity. However, if you throw the flashlight away from the ship, using an internal force from your muscles, the flashlight will acquire some net momentum away from the ship. Since total momentum is conserved (given the lack of external forces), your body needs to respond by acquiring momentum in the direction of the ship. Another way to think about this is that your hand will use the flashlight as a surface against which to push. As you push against the flashlight, it pushes back against your hand, providing a force on your body. The force is not as strong as it would be if you were pushing against an immovable wall because the flashlight can only resist your throw to a degree that corresponds to its inertia, which comes from its small mass. But it will do.

Another question remains to be solved: How long will it take you to travel the 10 m? That will depend on your mass, the mass of the flashlight, and the velocity that you gave to the flashlight. If we set this problem up carefully within the framework of conservation of momentum, we can calculate the time. If you happened to be working with a deep-space choreographer who has instructed you to make the trip in four counts, you will need to do the calculations to figure out how hard to throw the flashlight. (And the choreographer needs to give you a quantified description of a count!)

Let us set the stage. First, we can adopt the reference frame of the spaceship: an object traveling with zero velocity is an object that is not moving with respect to the ship. We will put the $x = y = 0$ position at the center of your mass, where you include your body, your spacesuit, and the flashlight that you are holding as your mass. We now need to choose how your axes will be aligned with the physical situation. For the purposes of this thought exercise, we only care about movement along the line between you and the ship. We can set your axes such that the x-axis is defined as the line directly between your center of mass and the center of mass of the spaceship.

We have already determined that $x = 0$ at your center of mass, but since we will be working with vector quantities, we need to pay careful attention to direction. Let's set the positive x direction as pointing from your center of mass to the ship and the negative x direction as pointing from your center of mass and away from the ship. This means that if you were to move in the positive x direction you would be moving toward the ship.

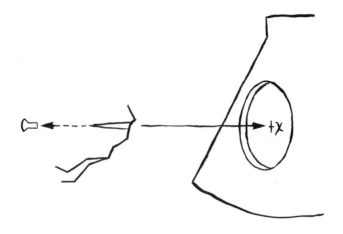

A few final variable assignments are needed. We will have two objects to keep track of: you (for which we can use the variable D, for *dancer*) and the flashlight (for which we can use the variable L for *light*, avoiding the confusion that might come from choosing F, which could mistakenly be associated with the word *force*). We are going to be working with momenta, and therefore masses and velocities for each of our objects. This is a conservation problem: we will be comparing some set of initial quantities with a set of final quantities. We will assume that momentum remains the same between these two snapshots in time. Two objects (D and L) with three quantities each (p, m, and v) in two snapshots of time (initial and final) gives $2 \times 3 \times 2 = 12$ variables. We can make one assumption to simplify the calculation: the masses of both you and the flashlight remain constant over the course of the problem. So instead of calculating using the initial and final masses of your body and the initial and final masses of the flashlight, we only need one mass for each, reducing the variables to ten.

p_{Di} : your initial momentum
v_{Di} : your initial velocity
p_{Df} : your final momentum
v_{Df} : your final velocity
m_D : your mass
p_{Li} : the flashlight's initial momentum
v_{Li} : the flashlight's initial velocity
p_{Lf} : the flashlight's final momentum
v_{Lf} : the flashlight's final velocity
m_L : the flashlight's mass

The total initial momentum (p_i) of the system that includes you and your flashlight is equal to your total initial momentum (p_{Di}) plus the initial momentum of your flashlight (p_{Li}):

$$p_i = p_{Di} + p_{Li} = (m_D v_{Di}) + (m_L v_{Li}) \tag{39}$$

Initially you and your flashlight each have zero velocity ($v_{Di} = v_{Li} = 0$). If we insert this information into the previous equation we will calculate that the initial momentum of this system is 0.

$$p_i = 0 \tag{40}$$

Given that we do not have external forces, we know that momentum will be conserved: the final momentum will be equal to the initial momentum

$$p_i = p_f = 0 \tag{41}$$

What is the expression for the final momentum?

$$p_f = (m_D v_{Df}) + (m_L v_{Lf}) \tag{42}$$

Thanks to conservation, because the initial momentum is equal to 0 we know that the final momentum must also be equal to 0:

$$p_f = 0 \tag{43}$$

Plugging in the expression for final momentum and setting it equal to 0 gives:

$$(m_D v_{Df}) + (m_L v_{Lf}) = 0 \tag{44}$$

We can write that equation in a slightly different way, by subtracting the second term from both sides of the equation:

$$m_D v_{Df} = -(m_L v_{Lf}) \tag{45}$$

The new equation shows us that the final momentum of the dancer is equal and opposite to the final momentum of the flashlight. You can plug in your mass, the mass of the flashlight, and the velocity that you want to achieve on your way back to the ship. The only unknown variable that remains is the velocity of the flashlight, which you can now calculate.

You can tell your deep-space choreographer that you can now calculate the flashlight velocity needed to return you to your ship in whatever amount of time he or she wishes.

Of course, being able to calculate velocity and time does not necessarily ensure that you can physically perfect the modulation of your velocity. You get only one chance here—once you have thrown your flashlight into outer space, it's gone. And if you get the angle of that throw wrong, you might miss your ship and send yourself off into oblivion. How does this performance end? Only you, your flashlight, and your math skills can answer that.

The Momentum of Two (or Three)

We would be hard-pressed to continue this book if we sent our readers flying off into the cosmos. So let's assume that you nailed your flashlight throw. You have, after all, been studying physics and dance for some time.

With your repertoire of embodied and quantitative know-how, you got yourself back to your ship and home safely to planet earth, where you are eager to continue your research into momentum with multiple bodies. Although you had grown attached to that trusty flashlight, it was an object, after all, not as warm-blooded or physically responsive as your dancing friends on earth. You reserve studio time and invite a group of dancers you admire to participate in your physics experiments.

If you try the previous experiment with a partner, the first thing you will discover is that flinging another person away from you does not have the same effect on earth as it would in outer space. The person's body does not go very far, and you barely move. Why is this?

The answer, of course, is that you have a rather large elephant in the room with you that was not present in space: our planet, that hulking, quiet, omnipresent additional partner in every dance. Earth is exerting external forces on your dancers, pulling them down by means of gravity and stopping their motion on its surface by means of friction. The earth complicates the physics of modeling conservation of momentum. But the concept of momentum is still useful.

What if we now turned the choreographic scenario inside out and asked you to investigate the effects of dancers throwing themselves *at*, instead of *away from*, each other? In the 1970s, the dance artist Steve Paxton researched such a phenomenon with a group of young dancers at Oberlin College. Inventing a new way of moving together, they emphasized the relationships among human beings in concert with natural forces. Weight, gravity, momentum, and torque were their topics of research. The dancers began simply, by teaching themselves to bear and move with the weight of other dancers— even those with larger mass—through skillful maneuvering in relation to natural forces: softening their bodies on impact, for instance, rather than stiffening. Or turning the momentum from the impact with another body into something else, such as a release to the ground, a rotation, or a roll. Instead of assisting each other exclusively with their hands, as in much of classical ballet, they made the surface of their entire body available for partnering and support. The dance form they created became known as *contact improvisation*.[10]

A 1972 film of dancers engaged in contact improvisation vividly illustrates their skillful play with momentum. The dancers experiment with flinging themselves through the air toward each other. In one segment, a sturdy woman in shorts is the catcher. Different dancers attempt the maneuver. One man jumps into her sideways, making contact at her midriff, curled up like a baby. She buckles to the ground under his weight. Deciding that something about that contact did not feel or look quite right, they try again.[11]

As the exercise evolves, the catcher learns that she can buffer the jumper's weight and her own fall by turning her torso along with the incoming dancer's momentum. Instead of colliding frontally, in other words, she rotates at the waist. She is then able to absorb the leaping dancer's momentum by riding the forward motion along with him.

This knowledge is put into immediate action with a new trial: a tall man launches himself at her, jumping high enough to make contact with his knee against her chest. As he reaches her chest, she grabs on to his body and twists at the waist, hooking into the arc of his leap. In contact, they ride the force of his jump together down to the ground.

You can find other models for partnering in dance—for example, one in which a catcher either resists and opposes the jumper's force, or assists in increasing his or her height and velocity. But the woman doing the catching in this film does not alter the incoming force. Instead, she figures out how to ride its effects alongside the other body. The unique beauty of contact improvisation lies in this attention to working together, both with other dancers and with the natural forces as they are generated.

Think of the jumping exercise captured in the film as practical experimentation with momentum. The main research question is: How can you absorb another's momentum fluidly and supportively? Knowing what we know about conservation of momentum in outer space allows us to note that the earth is another major factor in this scenario. For not only does the dancer doing the catching absorb the jumper's momentum—the earth does as well. This is a dance for three. Contact improvisation would work very differently in outer space.

This example reveals another dimension of momentum, considered in the context of human interaction. You have choices when you dance with another person: Will you accompany the momentum of another? Or will you block it? In dance, where all action is meaningful and interpretable, these two options imply different political stances, as much as physics calculations. Accompanying the momentum of another so the two of you generate that momentum together builds physical trust and support, qualities that can seem all too rare in our digitized age.

6. Turning

Bodies throughout the universe—whether celestial or dancing on earth—turn. Pairs of nearby black holes reach out to each other through their strong gravitational fields and spiral together to merge. Galaxies turn, as do the solar systems within them. Planets rotate about stars; moons rotate about planets. Our own moon has lost rotational energy over its 4-billion-year duet with our planet: with one side locked in place facing the earth, it no longer rotates about its own axis. In our shared elliptical path around the sun, our moon has lost its pirouette.

The same physics principles that govern the rotation of celestial bodies—torque, moment of inertia, and angular momentum—also enable a dancer to turn. Depending on the technique, dancers may pirouette on the balls of their feet or toes, pivot on their heels, or spin on the crowns of their heads. A dancer could take off and execute multiple 360s in the air, or dive-roll across the stage. Ever greater control over the natural forces comes through practice. One Guinness World Record breaker, the Japanese dancer Aichi Ōno, kept topping his own spinning record, which at last check hit 142 headspins in one minute. Using the top of his head as his pivot point, Ōno maneuvers his legs and arms into different configurations—elongated to the sky in one instant, curled up around his ears the next—thereby speeding himself up or slowing himself down. His headspins look like shape-shifting windmills in high winds, fixed and parallel to the earth. Ōno describes his mastery as a physical sensation: "I just feel the center point, running through my body, and I just concentrate on that point."[12]

Ōno's feats are the result of ardent movement research. For as the (much tamer) exercises in this chapter will demonstrate, turning in dance represents a sophisticated investigation of rotational motion. Technical rules are codified and handed down through inherited traditions and training, as in Ōno's immersion in breakdancing form. But they are also developed through trial-and-error investigation in the dance studio, in the same way that scientists experiment in the laboratory. The smallest technical adjustments to a dancer's turn can have a major impact on its physical efficiency. And just as dancers and choreographers engage in movement research, physicists have spent centuries researching everything from the circling of the planets to the circling of our bodies in space—eyeing the minutest wobbles along an elliptical path for insight into cosmic mechanics.

In this chapter, we link the concepts of torque, moment of inertia, and angular momentum to a dance history lesson on the evolution of the pirouette by the choreographer George Balanchine. Another expert movement researcher, Balanchine altered his dancers' engagement with physical forces by making subtle and not-so-subtle changes to Russian classical ballet technique, including the execution of a turn. His adjustments created startlingly new effects, even as they also produced a pirouette that required his dancers to exert less force to achieve the same speed in the turn. As we will see, it was as if the choreographer had been dreaming about physics all along.

The Pirouette

To trace the evolving mechanics of the pirouette, we need to travel back in time to 1913, the year that Balanchine began his studies at the Imperial Theatre School in Saint Petersburg, Russia. The institution had deep roots: Empress Anna Ioannovna had founded the school in 1738 as part of the establishment of a military academy. Nearly one hundred years later, Nicholas I—a balletomane—moved the school to its home on Theatre Street. By the early twentieth century, when Balanchine arrived, the school still retained a military feel.[13] The students' regimented daily training in classical ballet instilled in their bodies a Russian dance tradition that had been honed over centuries.

The ballet technique that Balanchine learned had five positions of the feet: first through fifth. The position most commonly used in the preparation for a pirouette is fourth.

A simple exercise will give you the feel of fourth position. First, face 45 degrees to your chosen front, and stand with your left leg in front and your right leg behind. This angle is called croisé, because the legs are crossed. Your feet should be slightly turned out—only insofar as you are comfortable, more or less depending on your familiarity with ballet training. In a pirouette en dehors, the turn moves in the direction away from the leg in front. In ballet, this is the "supporting leg," the leg engaged with the floor that serves as the axis of rotation. In your fourth position, your left foot is in front, so for an en dehors turn, your left leg becomes the supporting leg, and you will turn toward your right shoulder.

In Russian classical ballet, the pirouette en dehors starts in a plié in fourth position in which the front and back legs are both bent. In order to accomplish the turn, the dancer pushes into the floor with the back right foot, rises onto the balls of the feet or pointe, and shifts her or his weight over the supporting leg. The force to make the turn comes from the pressure of the right foot into the floor and the action of the arms, which we will describe in greater detail shortly.

Rotational Motion

To understand rotational motion in physics—in particular, the pirouette—we begin with some definitions. First, we must define the rotating mass under analysis. The mass that we will consider in this chapter is a dancer's body.

Next, about *what* will the dancer rotate? It will not be enough for us to define a point about which the dancer turns—you can turn your body a number of ways about a point in space. We need to define the *axis of rotation* about which the dancer will turn to sufficiently constrain the motion. For the pirouette, the axis of rotation will be the imagined vertical line that passes through the dancer's body, entering at the head and exiting at the point about which the dancer's toes pivot on the floor.

The organization of the body about this axis of rotation defines its *moment of inertia*, which determines the dancer's resistance to the turn. The more the dancer's body hugs the axis of rotation the faster the dancer can turn for a given push. As the dancer's mass extends out from the axis—through limbs stretching out, for example—the dancer's resistance to the turn increases. The moment of inertia quantifies a mass's resistance to turning.

And what sets the turn in motion? For this we will need to know not only the *amount* of force that is applied, but also *how* that force is applied with respect to the axis of rotation. The variable that elegantly pulls all of this information together is known as *torque*.

Finally, just as we had defined momentum as mass times velocity, we can define *angular momentum* as moment of inertia times angular velocity. Both definitions of momentum combine an object's resistance to motion with the speed of the motion, and both are conserved in the absence of external forces.

The Lunge

Fast forward in time: in the 1930s, Balanchine immigrated to America with the help of the philanthropist Lincoln Kirstein. Balanchine had already acquired a reputation in Europe as a significant new talent in ballet through his work as a choreographer for Sergei Diaghilev's Ballets Russes. Kirstein aimed to create an American ballet, and he saw Balanchine as the artist who could make it happen. In 1934, they created the School of American Ballet to begin training high-caliber young dancers. Fourteen years later, in 1948, they officially established the New York City Ballet, a company that became Balanchine's laboratory for aesthetic innovation.

Balanchine constantly experimented with form. He sped up tempos, insisted on stretched, elongated lines and quicker petite allegro (small quick jumps and movements), and emphasized the *transitions* between positions as much as the positions themselves. He removed what he considered to be static nineteenth-century pantomime, and replaced conventional narrative form with abstract imagery based on movement for movement's sake. Port de bras—the movement of the arms—became more fluid and closer to the body, and the ballerina's feet became stronger and more sculpted. (In one of his steps, the ballerina is instructed to present the foot like an elephant's trunk unfurling.) He also built surprises into his technique: jumps and turns occurred unexpectedly under his choreographic direction. While his alterations reflected the excitement and energy of the American culture that he observed outside of the theater, these changes emerged through intensive time spent *inside*—working in the studio with his dancers, researching ballet technique.

Balanchine's preeminent muse, Suzanne Farrell, vividly describes the day that he altered the pirouette. He first zeroed in on the fourth position used to prepare for the turn. As she notes, up until that day she and "every other ballet dancer on earth" had been trained to prepare with both legs bent in plié. But that morning Balanchine coaxed her into a longer, deeper lunge. With her back right leg straight and her weight shifted over her front supporting leg, she moved her legs farther apart, and farther apart again. "Bigger," he insisted—a request he made frequently when coaching his dancers. Thinking she was being set up for failure, Farrell managed a lunge so low and deep that she nearly extended into a split, and then, at Balenchine's instruction, she took off. She turned and turned ... It was, she writes, "the most glorious" pirouette she had ever felt.[14]

In making these adjustments to pirouette technique, Balanchine might as well have been a physicist armed with formulas and careful calculations. His pirouette preparation gave dancers greater bang for their buck: less force was required to execute multiple turns. And with the weight already forward over the supporting leg, the dancer need not move forward and up, but simply up, into relevé on pointe. The new preparation also lent a whiff of surprise: from the classical fourth position with knees bent, the audience knew what to expect. The dancer was clearly going to turn. In contrast, from

the deeper lunge the dancer might move forward, backward, or side to side. He or she might jump, travel across the stage—or pirouette.

By extending the fourth position into a lunge, Balanchine had in fact increased the distance between the two points of support—the right and left legs. How this increase in distance affected the turn connects the pirouette to the physics concept of *torque*.[15]

Torque

The relationship between the speed of a turn and the force applied to create the turn can be explored through the concept of torque. We can think of torque as analogous to force, except it produces what is called an *angular* acceleration instead of a linear acceleration. In other words, the dancer is spinning faster instead of speeding up along a line.

In order to define torque, we need to incorporate a few variables into our description of the physical situation. When you pivot on one foot (we will call it foot *A*) by pushing off with the other foot (which we will call foot *B*), you can control many variables in order to maximize the initial speed of your turn. The most obvious variable is the force with which you push into the ground. The maximum force that you can achieve is related to your strength, balance, and experience. The *direction* of the applied force is also important. We can illustrate this through a diagram of the two feet with a line connecting them, in which the point marked on foot *A* can be considered the pivot point for the turn, and the point marked on foot *B* can be considered the location from which the force for the turn is being supplied.

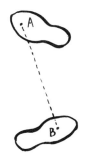

If you push off with foot *B* in a direction directly toward foot *A*, which is along the *line of action* connecting the two feet, you will move forward rather than spin. If you push off in a direction that is perpendicular to the line of action, you will maximize the speed of your turn.

To start the turn, foot *B* pushes on the ground with force *F* as shown in the drawing that follows. In accordance with Newton's 3rd Law of Motion, the ground pushes back on foot *B* with an equal force in the opposite direction, which results in a turn in a clockwise rotation (en dehors) when viewed from above. Forces that are applied between the perpendicular and parallel directions can result in a turn, but you can maximize your turn speed by applying the force in a direction that is perpendicular as shown in the drawing. Try this exercise, with forces applied parallel, perpendicular, and somewhere between the two to get a sense of the impact of the direction of the applied force on your rotational motion.

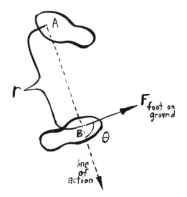

There is another variable that you can control that has a direct impact on the initial speed of your turn, and that is the distance between foot *A* and foot B. If your feet are placed very close to each other, so that the length of the line that connects foot A and foot B is short, more force needs to be applied in order to begin with high angular velocity. If you move your feet farther apart, it becomes easier to spin with high angular velocity until you reach a point where it becomes difficult to balance or you can no longer effectively apply the force. Imagine you have a wrench and want to unstick a bolt. The longer the wrench handle, the more turning force you can apply for your arm strength. When Balanchine instructed his dancers to extend their lunge, he was similarly increasing their turning force for a given push off from the ground.

If we restrict ourselves to the lengths between point A and point B over which you are able to effectively apply a turning force with foot B, we can examine the start of the turn in the context of *torque*, for which we use the symbol τ. Its formula is

$$\tau = rF \sin \theta \tag{46}$$

where r is the distance between the pivot point and the applied force F, which we can consider to be under the balls of feet A and B as marked in the previous diagrams, and the angle θ describes the direction of the application of the force with respect to the line of action. The units of torque are units of distance multiplied by units of force, or meters times newtons (mN).

Note that the sine of an angle will equal a number between −1 and 1, and that the sine of 90 degrees equals 1. This means that when the angle θ is equal to 90 degrees, the formula becomes

$$\tau = rF \sin 90 = rF(1) = rF \tag{47}$$

The number 1 is the maximum value that the sine of an angle can take, so we can see from the formula that one way to maximize the torque is to apply the turning force at a 90-degree angle to the line of action. Other ways to provide the maximum torque, according to the formula, are to have a greater separation r between the two feet, as Balanchine engineered through the deep lunge, and to maximize the force F that the dancer applies to initiate the rotation.

Of course, this only works within a range of values of r over which a force can efficiently be applied that results in a turn. We are also assuming that the turning body is rigid when we model a pirouette with the formula for torque. In order to understand why this is necessary, let us return to the case of a force applied at the end of a wrench to turn a bolt. If your wrench is made of steel, the force that you apply to the wrench is transferred to the bolt. If, however, your wrench were constructed out of soft clay, applying a turning force to the wrench would merely fold the clay. It is probably safe to assume that the human body acts more like steel than clay in this instance, but it is important to keep in mind that the body is complicated. Sometimes when we model the motion of the body with simple formulas, we are ignoring potentially important details.

It is part of the richness of the study of physics that we can adjust our model to have more accurate predictive power by taking into account more of the details of the physical reality. But it is also a limitation of using physics to understand dance that capturing the details of a typical body, to say nothing of the relevant characteristics of an individual, quickly becomes complicated. At those moments, we find movement research may more quickly lead us to an understanding of the laws of nature than our exploration of mathematical models.

The Arms

After asking his dancers to start their turns in a wide lunging fourth position, Balanchine made other changes to the pirouette as well. He trained his dancers to "spot" front—that is, whip their heads around to focus their eyes on a single spot above the heads of the spectators—thus presenting the pirouette to the audience, rather than to the corner of the stage. He also noticed an incongruity that needed to be tweaked: the conventional starting position of the arms no longer matched the sleeker, athletic lines of the legs.

Russian classical ballet technique codifies five positions of the arms—first through fifth, similar to the positions of the feet. Balanchine had been trained to use third position of the arms as the starting position for the pirouette. In third position, both arms are rounded, with one arm curved more, bisecting the front of the body, and the other extended outward from the shoulder to the side.

Rather than curved arms, Balanchine asked his dancers to use an elongated port de bras, in which both arms dramatically extended away from the body. Remember for a moment the fourth position of the feet that you tried earlier in this chapter. The arm reaching forward would be your right arm, and your left arm would reach straight out from your side. Your right arm, reaching along the diagonal, mirrors the line of the back right leg in the deep lunge.

Balanchine used poetic imagery throughout his teaching and choreographic practice, and the final touch to this position is the image he used to describe the elongation of the arm on the diagonal: "reaching for diamonds." (The teachers at the School of American Ballet sometimes replaced this idea with "reaching for chocolate"—to each his or her own desire.) The different positions can be seen in the photographs and illustrations.

In altering the starting placement of the arms, Balanchine was tapping directly into another key physics concept: *moment of inertia*.

Moment of Inertia

Just as we defined mass with Newton's 2nd Law of Motion, with respect to the amount of acceleration an object would experience under an applied net external force, we can define an object's moment of inertia in relation to the amount of angular acceleration an object would experience under an applied net external torque. The two analogous equations are shown below:

$$F = ma \quad \text{Newton's 2nd Law for Linear Motion}$$
$$\tau = I\alpha \quad \text{Newton's 2nd Law for Rotational Motion}$$

Here we introduce the new variables I for moment of inertia and α for angular acceleration. If you apply a force to a small mass you will get a larger linear acceleration than if you applied that same force to a large mass. Similarly, if you apply torque to an object with a small moment of inertia you will get a larger angular acceleration than if you applied that same torque to an object with a large moment of inertia. We can therefore think about an object's moment of inertia as its ability to resist rotation.

To calculate the moment of inertia, it is essential to first identify the axis of rotation. It is also necessary to know the distribution of an object's mass about the axis of rotation. As a simple example let's consider a group of several masses that have known values and exist at well-defined points in space. In this scenario, the formula for moment of inertia about a rotational axis is

$$I = m_1 r_1^2 + m_2 r_2^2 + m_3 r_3^2 + \ldots \tag{48}$$

where each object is indexed (see subscripts), from one to the total number of objects n, and the mass m and distance from the axis of rotation r are considered in the calculation. Let's take the example of a system of three masses—m_1, m_2, and m_3—that are located on the x-axis at –2.0 m, 1.0 m, and 1.5 m, respectively. If we choose our axis of rotation to be a line that passes through the point $x = 0$ and is perpendicular to the x-y plane, our values of r_1, r_2, and r_3 are –2.0 m, 1.0 m, and 1.5 m, respectively. If we know the masses of the three objects, and if we connect them through a thin, massless rod along the x-axis, we can calculate the moment of inertia of the system of masses. For simplicity, let us assume that each of our three objects has a mass of 1 kg. The calculation is then

$$\begin{aligned} I = & \ (1 \text{ kg} \times (-2.0 \text{ m})^2) + (1 \text{ kg} \times (1.0 \text{ m})^2) + (1 \text{ kg} \times (1.5 \text{ m})^2) \\ = & \ 4 \text{ kg m}^2 + 1 \text{ kg m}^2 + 2.25 \text{ kg m}^2 = 7.25 \text{ kg m}^2 \end{aligned} \tag{49}$$

As elegant as that calculation is, a dancer who is turning does not have too much in common with a system of three masses along the x-axis connected by a massless rod. How can we gain insight into a realistic physical system using our formula for moment of inertia? In particular, we would like to understand how a person's moment of inertia about an axis of rotation changes as his or her arms or legs move with respect to the axis.

We might imagine modeling the dancer as a series of connected, discrete masses. If we think about the dancer as a series of connected cubes we could picture dividing the mass of the dancer into the number of 1 cm or smaller cubes that would make up the body, and we could measure the dimensions of the person to understand how far each of our cubes is from the axis of rotation. We could take the center of each cube as the location of that cube's mass when we calculate the r values for our formula. This is a tremendous amount of work that is valid only as long as the dancer does not change his or her position with respect to the axis of rotation. And we could do a better job if we did not assume that the person had uniform density because we know that organs, blood, bones, muscle, and fat all have different densities.

In order to explore Balanchine's change to pirouette technique through a physics calculation, we will focus on the positions of just one arm. Even this simplification is complicated, but it does offer some insight. We will calculate the moment of inertia of the right arm in Balanchine's technique and leave the calculation for the classical Russian technique as an exercise in the workbook.

To simplify the calculations, we will use information about average dimensions and densities for young adult women and men from *The Neuromechanics of Human Movement* by Roger Enoka, with relevant excerpts shown in the table below.[16] We will consider the arm as a system of these three masses: the hand, the forearm, and the upper arm. By using data about the locations of the center of mass of these body parts, we can see how their locations change between the two positions and therefore understand the difference in initial moment of inertia between the two techniques. For simplicity, we will assume that the width of the torso is 30 cm.

Segment	Length in cm for young adult women (men)	Mass percentage of total body mass for young adult women (men)	Center of mass location % along length for young adult women (men)
Hand	7.80 (8.62)	0.56 (0.61)	25.26 (21.00)
Forearm	26.43 (26.89)	1.38 (1.62)	54.41 (54.26)
Upper arm	27.51 (28.17)	2.55 (2.71)	42.46 (42.28)

The first line of this table tells us that the average length of the women's hands measured was 7.80 cm. Each hand was on average 0.56% of the total mass of the body. So if the women had a total mass of 100 kg, a typical mass to expect for a hand would be 0.56 kg. The center of mass of their hands was much closer to their wrists than to their fingertips, only about a quarter of the length (25%) from the base of the hand. If the center of mass location were halfway from each end the percentage would be 50%.

In the following diagram, the upper arm, forearm, and hand are outstretched in a straight line perpendicular to the plane defined by the dancer's chest in the Balanchine technique. In reality, we would expect the angle between the torso and the right arm to be more obtuse, to reflect the dancer's extreme reach, but this simplification will make the calculation easier. In this diagram the axis of rotation passes through the center of the torso. The body segments are labeled with dimensions for an average woman from the table:

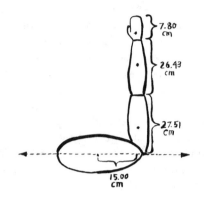

In order to do the calculation, we need to know the distance from the centers of mass of each segment of the arm to the axis of rotation. We therefore need to know not only the length of each segment of the arm but also the locations of their centers of mass. We can do this in a few steps. We will first calculate the distance from the base of each body segment (hand, forearm, upper arm) to the center of mass position of that body segment.

For the hand:
$$7.80 \text{ cm} \times 25.26 = 1.97 \text{ cm} \tag{50}$$

For the forearm:
$$26.43 \text{ cm} \times 54.41 = 14.38 \text{ cm} \tag{51}$$

For the upper arm:
$$27.51 \text{ cm} \times 42.46 = 11.68 \text{ cm} \tag{52}$$

Adding this information to the diagram, we have:

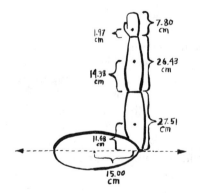

This still does not tell us the distance from the segments' center of mass positions to the rotational axis. For this, we can form a series of triangles. The base of the triangle will be the 15.00 cm length from the axis to the perpendicular line that follows the outstretched arm. Then we will need to find the distance from that base to each center of mass position. For the upper arm this is simplest: the distance to the upper arm's center of mass is the number we calculated, 11.68 cm. But for the forearm, the center of mass position with respect to the base is the the one we calculated (14.38 cm) plus the entire length of the upper arm (27.51 cm). The hand's center of mass is farthest away from the axis of rotation. We can use these positions to form three right triangles, where one side is the base length of 15.00 cm, another side is the distance from the line defining the base to the center of mass location, and the hypotenuse gives the actual distance from the axis of rotation (r).

A useful formula from geometry tells us that the length of the hypotenuse of a triangle can be found in this way:

$$hypotenuse = \sqrt{(Base_1)^2 + (Base_2)^2} \qquad (53)$$

where the other two sides of the triangle are $Base_1$ and $Base_2$.

Confirm the following dimensions:

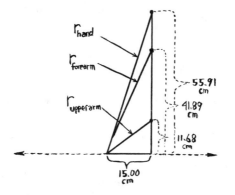

$$r_{Hand} = \sqrt{(15.00 \text{ cm})^2 + (55.91 \text{ cm})^2} = 57.89 \text{ cm} = 0.5789 \text{ m} \qquad (54)$$

$$r_{Forearm} = \sqrt{(15.00 \text{ cm})^2 + (41.89 \text{ cm})^2} = 44.50 \text{ cm} = 0.4450 \text{ m} \qquad (55)$$

$$r_{Upperarm} = \sqrt{(15.00 \text{ cm})^2 + (11.68 \text{ cm})^2} = 19.01 \text{ cm} = 0.1901 \text{ m} \qquad (56)$$

We now have the distances between each of our arm segments and the axis of rotation. In order to calculate the contribution of an arm to the moment of inertia, we need to assume a mass for the dancer we are modeling. If we consider a woman with mass of 60 kg, we can use the information from the table to estimate the mass of each segment. Note that the table gives values of mass fraction in the form of percentages. If a hand is typically about 0.56% of the mass of the entire person, we can calculate the mass of the hand by multiplying the entire mass by the value 0.0056.

$$m_{Hand} = (60 \text{ kg}) \times (0.0056) = 0.34 \text{ kg} \tag{57}$$

$$m_{Forearm} = (60 \text{ kg}) \times (0.0138) = 0.84 \text{ kg} \tag{58}$$

$$m_{Upperarm} = (60 \text{ kg}) \times (0.0255) = 1.53 \text{ kg} \tag{59}$$

The moment of inertia contribution of the right arm is therefore approximately 0.34 kg m^2 for a 60 kg woman in the position designed by Balanchine, as shown below:

$$
\begin{aligned}
I &= I_{Hand} + I_{Forearm} + I_{Upperarm} \\
&= (0.34 \text{ kg})(0.5789 \text{ m})^2 + (0.84 \text{ kg})(0.4450 \text{ m})^2 + (1.53 \text{ kg})(0.1901 \text{ m})^2 \\
&= 0.1139 \text{ kg m}^2 + 0.1663 \text{ kg m}^2 + 0.0553 \text{ kg m}^2 = 0.3355 \text{ kg m}^2 \tag{60}
\end{aligned}
$$

We have included an exercise for calculating the beginning position for the classical Russian technique in the workbook. Since the mass of each section of the arm does not change between the two calculations and the distance from the center of mass of each section to the axis of rotation either stays the same or decreases between the Balanchine technique and the classical Russian technique, we expect the moment of inertia calculated for the Russian technique to be less than the value calculated for the Balanchine technique. Verify that this is true once you have finished your calculation.

The Turn

One more detail about the Balanchine pirouette remains to be covered that relates directly to moment of inertia: the position of the arms during the execution of the turn. Think back to the headspins of Aichi Ōno that we described earlier, and the way that Ōno's redistributions of his limbs around his axis of rotation gave him greater control over the speed of his rotation. As we will see, Balanchine similarly rearranged the mass of the dancers' arms during the pirouette, thereby reducing their moment of inertia.

His change was deceptively simple. In Russian classical ballet training, the arms resolve from the starting third position into first position, in which both arms round to form a circle in front of the body. Notice this position of the arms in the figure on the left in the drawing that follows.

Balanchine changed this important detail. Instead of turning with the arms away from the body, he had his dancers fold their arms in close to the body. As usual, he preserved enough classical form to be recognizably balletic, while significantly altering the details: the elbows bend lower in toward the dancer's sides, the forearms face the body, and the wrists cross. You can see this altered position in the figure on the right.

Now you try. Assume the fourth position, with your left leg forward. Your weight should be fully positioned over your front left leg in plié. You will want to press into the floor with your right foot, to supply the force for the turn. All you need to do from this starting position is relevé straight up. In this case, because you are most likely not wearing pointe shoes, you will rise only onto the ball of your foot. Notice that instead of shifting your weight from a center of mass located between your two legs, as you would in Russian classical ballet technique, you have positioned your center of mass over the front supporting leg, right where you need it to be in the relevé. As you rise, pull up your right toe to touch the left knee, with the right knee turned out to the side. In ballet, this position is called passé. One more detail: as soon as you relevé, pull your arms in so they are close to your body, as in the figure on the right in the drawing. The combination of the right foot pressing into the floor and the right arm pulling in quickly provides the force you need to take you around.

To be sure, there are many more technical details here than anyone can master from reading this book. For a more basic exercise, you can simply try turning in place. First hold the arms extended away from the body while turning, and then pull them in close to the torso. You can feel the effects on moment of inertia upon your speed of rotation—which, in physics terms, is your angular velocity. Even this basic exercise will give you some sense of the nuanced relationship between Balanchine's choreographic practice and the forces at work on dancers' bodies.

Angular Momentum

One more useful tool in understanding Balanchine's design for the pirouette is the concept of angular momentum. We defined linear momentum p as a measure of the mass of an object m times its velocity v. In a similar way, angular momentum is the measure of the moment of inertia of an object

I times its angular velocity ω. The equation for angular momentum L is therefore

$$L = I\omega \tag{61}$$

Like linear momentum, angular momentum is conserved in the absence of external forces. Think about what happens to the angular velocity of a turn when an object's angular momentum remains constant as its moment of inertia changes.

In a scenario with very little friction, such as an ice skater performing a turn, you can imagine what happens when the skater pulls in her or his arms or legs closer to the rotational axis. It has the effect of bringing mass in closer to the axis of rotation, which lowers the values of r for the limbs that are moving and results in a lower I. Because I decreases and L is conserved, ω must increase, and this is exactly what happens. The spin velocity goes up when the arms or legs come in and goes back down when the arms or legs are extended. Although a dancer doing a turn has significant friction with the floor through the foot, so external forces are at work and angular momentum is not conserved, the impact of moment of inertia changes is still relevant. The Balanchine technique's increased moment of inertia before the turn and reduced moment of inertia during the turn, when compared with the classical Russian ballet pirouette, can lead to higher angular velocity for the given amount of torque.

Kinesthetic Intelligence

Not all dancers turn with ease and fluency. Those who do are known as "natural turners." Suzanne Farrell was a natural turner in ballet, just as Aichi Ōno is in breakdancing. They share an intuitive understanding of rotational motion, even as the details of their physical placement differ significantly, as dictated by their dance forms. Theirs is a unique form of kinesthetic intelligence.

The Lithuanian dancer Lora Juodkaite has cultivated another approach—a practice of gyration, or the ability to spin for hours. The French-Algerian choreographer Rachid Ouramdane has featured her singular capabilities in his dances, including *Tordre* (2014), a dance portrait for two women, in which Juodkaite's spinning played a memorable role.

As Juodkaite turned in *Tordre*, she traveled in a circular pattern around the stage. Her lower body reflected ballet training: she rose onto the balls of her feet, keeping her legs long, and executed small, stepping half-turns that looked like chaînés—quick, sequential traveling turns in classical ballet. Unlike in ballet, however, she moved at a slower, more deliberate pace, maintaining a steady rhythm that she modulated by manipulating her arms into different configurations. She arched her arms over her head, like a bird, or clenched her fists in front of her, like a fighter. The spinning created a constancy against which these images surfaced and receded, even as they motored her rotation. Juodkaite did not spot, the typical dance technique trick used to avoid dizziness, which allowed her the freedom to look up, down, or straight ahead with no effect on her balance. Sometimes tighter and faster, sometimes opening up like a lazily revolving door, her patterns were always precise. As Juodkaite spun, she spoke. She explained that she had been developing her gyration practice since childhood, and that while turning she felt a supreme sense of comfort and peace.[17]

We could break down the physics of her feat in terms you now understand: the placement of her arms affected her moment of inertia, while her feet provided the torque, or force, to maintain the constant turning. But this would not fully describe the mysterious effect of watching a dancer with so high a degree of kinesthetic intelligence that it seemed she could rotate until the end of time.

Part II: Energy, Space, Time

7.   Energy

In 2010, the choreographer Ralph Lemon created a dance that included six expert dancers moving at a furious rate for twenty minutes. The performers tossed themselves wildly about the stage, drawing closer and closer to total exhaustion and disorientation. Their movements—convulsing, leaping, slapping, quaking—were neither set nor predetermined. Instead, they were a by-product of physically investigating the concept of *fury*.[18] In asking his performers to achieve altered states of energy at high velocity, Lemon wanted to create a dance without form or style, one that might barely even be perceived as dance.[19] The more energy the dancers expended, the more they seemed to generate. As their exhaustion intensified, their energy reverberated outward: the stage seemed to vibrate. The larger piece in which this improvisation appeared, titled *How Can You Stay in the House All Day and Not Go Anywhere?*, was a meditation on profound human loss. In researching forms of grieving, Lemon replaced conventional emotional expression onstage with all-out kinetic release.

In the early twentieth century, physicists began to conceive of energy in entirely new ways. Energy (E) becomes fundamentally entangled with mass (m) and the speed of light (c) in Einstein's famous equation $E = mc^2$. The ability to trade matter for energy, and vice versa, changed how we understood conservation of energy in our models of nature. Equally revolutionary was the discovery within quantum mechanics that energy is quantized: it takes on a limited number of values, which nature sets. Amounts that fall in between these values are simply not possible. Einstein's mass-energy equivalency and quantization become noticeable only under the extreme physical conditions of high energy or small spatial dimensions, conditions that are not often accessible to people. Still, when choreographers like Lemon shift the choreographic field from crafting shapes with the dancers' bodies to composing the energy they emit, mass and energy can appear to be one and the same.

In this chapter, we weave through a series of meditations on how energy is defined, imagined, and deployed in physics and dance. While we can by no means offer a comprehensive treatment of energy in either field, we cover a wide range of examples. The discipline of physics and the discipline of dance necessarily define energy differently, and even within each discipline the definitions vary widely. Physicists study many aspects of en-

ergy, from kinetic energy to the electromagnetic spectrum to dark energy. In dance, energy becomes visible when dancers move, but it is also felt by the audience, as when a performer changes the intensity in the theater through shifts in focus and attention.

If we generalize from the various forms, formulas, and definitions that characterize energy in physics and dance, we discover compelling shared territory: both disciplines recognize *energy as a capacity to cause change*. And both disciplines recognize that *energy can assume different forms*. As we move through the chapter, we will indicate other connections and analogies. But you might also find yourself noticing connections of your own. As you read, consider the elusive travels of energy across physics and dance: Where do the ideas connect? When must they diverge? And why?

Kinetic Energy

How would we quantify the energy of a dancer hurtling through space? Kinetic energy is the energy of motion, and it depends on both the dancer's velocity and his or her mass. We use the symbol *KE* for kinetic energy. Its definition is:

$$KE = \frac{1}{2}mv^2 \tag{62}$$

where the variable *m* is the mover's mass and the variable *v* is the mover's velocity. We intuitively understand that the energy of motion increases as the dancer moves more quickly and his or her velocity increases. The opposite happens when the dancer slows down: as velocity decreases, kinetic energy correspondingly decreases. It also seems reasonable that larger masses moving at a given speed would have more kinetic energy than smaller masses moving at that same speed.

This formula introduces a new quantitative measurement category for the concept of energy. In physics the questions that need to be answered are "How is energy defined in terms of quantities that we can measure? What are the units?" Energy has a dedicated unit, the Joule, named after the physicist James Joule. The abbreviation for Joules is a capital J. Joules depend on mass, distance, and time. They are based on kilograms, meters, and seconds, respectively:

$$1\,J = 1\,kg\,m^2/s^2 \qquad (63)$$

We could also write this in terms of newtons, the unit we use to measure force. Since a newton depends on the same fundamental units, the equation could be written as:

$$1\,J = 1\,N\,m \qquad (64)$$

As we see from this equation, we can convert from a force to energy by multiplying the force by a distance.

Energy as Form

When describing the limited range of choreographic options, the choreographer Merce Cunningham once said that there are only so many different arrangements of the human body's arms, legs, head, and feet. But dance not only organizes human anatomy, it also plays with kinetic energy. There are as many ways of marshaling kinetic energy as there are individual human beings on the planet.

Different dance forms can be read in terms of their play with kinetic energy. The newer street dance "flexing," for instance, evolved from Jamaican bruk-up and reggae music clubs in Brooklyn. Flex dancing incorporates certain specific features while also making room for each mover's distinct style. Fine gradations of energy occur: a dancer may slide through one sequence and then abruptly lock and redirect the action to the next. Movers add new dimensions to the form by changing the emphases, or carrying their bodies in idiosyncratic ways. One dancer may prefer to glide along with the feet, while another adeptly works a particular flexion in the arms and spine. Another flickers through the occasional balletic image—shadows of past training rearing up? Each flex dancer synthesizes distinct kinetic energies—from popping to contemporary dance—that drive his or her formal innovations. Flexing has moved onto the concert stage, and the change in venue adds another layer of complexity to its presentation.

Choreographers in other dance idioms also manipulate energy. Twyla Tharp synthesizes kinetic energies to create dances that burst across the stage. Tune your attention in just the right way when watching her work and the choreography transforms into pure force. Tharp's "The Golden Section," the finale of her evening-length work *The Catherine Wheel* (1981), is such a dance. Propelled by David Byrne's musical score and decked out in shiny gold costumes designed by Santo Loquasto, the dancers race through seemingly impossible moves: they leap through the air into each other's arms; they spin non-stop. An alchemist of motion, Tharp melds together diverse movement vocabularies, drawing on ballet, jazz, and tap, as well as non-dance forms such as aerobics, yoga, and basketball. The dancers' kinetic frenzy exudes optimism, even joy, in contrast to the dark family drama that comes in the piece's earlier sections.

Gravitational Potential Energy

Choreographers' compositional decisions are constrained by what positions and structures are physically possible, as well as by the forces acting on the dancer's body at any given moment in time. Physics relies on potential energy to quantify what might ultimately be possible for a specific object. If the dancer decides to move, this potential will then be translated into kinetic energy.

One of the fundamental, ever-present forms of potential energy is *gravitational potential energy*. The force of gravity provides a mechanism for humans to store energy that their bodies can later access. For example, when people stand up, they do work against gravity. As they increase the height of their center of mass above the surface of the earth, they have farther to fall, and thus the potential for greater velocity.

We could calculate changes in gravitational potential energy between two different relative positions of any objects that have mass, including the earth and the moon, other planets, or two dancers moving in the studio. The masses of the pair of dancers, however, are too small in gravitational terms for the dancers to sense each other's gravitational presence or experience any changes in their combined gravitational potential.

For now, we will limit our discussion to the changes in gravitational potential energy that we experience as moving bodies on the surface of the earth. By convention, we set our coordinate system such that gravity acts along the y-axis. Positive y (+y) points up into the sky and negative y (−y) points down toward the center of the earth. Since gravitational potential is useful in the context of changes in height from one moment of time to another, we can quite conveniently set the $y = 0$ horizontal plane to be anywhere. It often makes sense to use the beginning or end point of a motion as the $y = 0$ point. What is important is that we set the coordinate system before attempting a calculation and are consistent throughout the performance of the calculation.

What do we calculate? The formula for gravitational potential energy for a mass near the surface of the earth, for which we use the variable U_G, depends on:

- the mass, m.

- the acceleration due to gravity on the surface of the earth, g, which already contains information about the earth's mass and the distance from the surface of the earth to its center. As previously defined, g = 9.8 m/s^2.

- the change in the center of mass y position from the starting to the ending position. This is equal to the final height minus the initial height: $y_{final} - y_{initial}$. We refer to this change as h because it refers to a change of height.

Gravitational potential energy U_G is therefore:

$$U_G = mg(y_{final} - y_{initial}) = mgh \qquad (65)$$

Let's try a sample calculation.

What is the change in gravitational potential energy when a dancer stands up from a prone position on the floor? Let's assume that the dancer's mass is 80 kg and that his or her center of mass is raised by 0.75 m. For convenience, set the initial height, $y_{initial}$, at 0 m and the final height, y_{final}, at 0.75 m so that $h = (0.75 \text{ m} - 0.0 \text{ m}) = 0.75$ m. The formula is then:

$$U_G = mgh = 80 \text{ kg} \times 9.8 \text{ m/s}^2 \times 0.75 \text{ m} = 588 \text{ kg m}^2/\text{s}^2 = 588 \text{ J} \qquad (66)$$

What if, instead of standing up, the dancer moves in the opposite direction, and falls down? For the same mass and change in height, the only difference in our formula would be the swapping of the initial and final position.

$$h = y_{final} - y_{initial} = 0.0 \text{ m} - 0.75 \text{ m} = -0.75 \text{ m} \qquad (67)$$

$$U_G = mgh = -588 \text{ J} \qquad (68)$$

When the change in gravitational potential energy is positive, the dancer has done work against gravity. When it is negative, gravity has done work on the dancer. And note that it doesn't matter how the dancer got to a standing or lying position. He or she could have done five laps around the studio between the starting and stopping positions and the only thing that would be needed for the calculation are the starting and stopping heights. The total change in gravitational potential does not depend on the path taken to get between your initial and final positions.

Falling

Choreographers exploit the drama inherent in potential energy perhaps most of all when they ask their dancers to fall. There are many different ways to fall toward the earth: by rolling, dropping, tossing, giving in, tipping, and

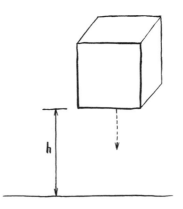

angling, among others. Two examples of falling—one in a ballet by George Balanchine and the other in a work of Tanztheater by Pina Bausch—reveal just how different "tipping" can seem in different contexts.

In the second movement of Balanchine's *Symphony in C*, the lead ballerina stands on her toes in profile to the audience, her arms raised overhead, and falls backward. Her partner first supports her with his hands at her waist and then lets go, with little fanfare save for gently whisking his arms outward, slightly away from her body. Her fall lasts less than one second, the space between two notes in Georges Bizet's musical score: 1...2. By "2" the ballerina's partner has caught her again. Certain in her physics experiment, the ballerina remains composed. She does not collapse: she trusts that her partner will catch her, and permits the physical forces acting upon her elongated body to speak.

In this pas de deux, as duets are termed in ballet, Balanchine introduces gravitational potential energy becoming kinetic for dramatic effect. Nothing about the ballerina's musical fall is out of control. However, even ballet cannot deny that without human resistance, gravity will pull a ballerina to the floor. Balanchine's choreography showcases a split second of natural anarchy within a controlled adagio—like a wildflower cropping up in an otherwise perfectly groomed landscape.

Another striking example of falling can be found in a duet from the piece *Nur Du* by Pina Bausch, which appears in the Wim Wenders film *Pina*. As in *Symphony in C*, it is a male-female duet, and the woman is the one who falls. Like the Balanchine ballerina, she holds her upper body straight and elongated, embodying classical form. Here ends the parallels between the two choreographies, however, for the Bausch dancer has just stridden across the pavement in heels and a floor-length gown. She now falls forward, instead of backward, and her partner catches her just before her face hits the ground. It is frightening to watch (and must be to perform)—but the woman appears to be too emotionally numbed to feel fear. With her arms held at her sides, shut off from the outside world, she slows to a standstill, hovers, and falls. Her partner in turn follows and catches her in the same spot each time, about a foot from harm. The pause before she tips unleashes a vicarious fear of falling: the viewer feels the transformation from potential to kinetic energy, with terrifying effect.

The fall in Balanchine's ballet is abstract, or storyless, and the dancers' vocabulary, coded interactions, and attire keep realism at a distant remove. His pas de deux hints at romantic love, but it is foremost an illustration of Bizet's music and classical ballet form. The Bausch fall, on the other hand, adds a measure of social context, with the female dancer's evening wear and face-first plunge. The duet alludes to the pas de deux form of ballet history, but the woman's repeated falls—and the psychological instability they signify—appear to be socially produced, a symptom of the couple's intimacy.

The depiction of falling does not only occur only within male-female partnerings. Women catch women, men catch men, individual dancers yield to the earth without a partner and find ways to rise again. The act of falling may seem straightforward across these scenarios. But in the hands of master choreographers such as Balanchine and Bausch, complicated dramatic meanings can emerge out of the simplest drop.

To experiment with the complexities of falling, craft a movement phrase that capitalizes on the concept of gravitational potential energy. Your phrase may have moments of suspension, hovering, or "air" time, angled to the floor. Focus only on potential energy as it transforms into kinetic energy. Remember as you experiment that even walking is a kind of falling. What kind of drama might you discover if you break down the movement's inception and execution? If you plan to investigate the dynamics of falling to the floor, have a partner to catch you or mats to cushion the fall.

Spring Potential Energy

Gravitational potential energy has to do with a dancer's placement in space relative to the earth: the dancer's potential energy due to gravity becomes kinetic energy as she or he begins to fall. There are other stores of potential energy that a choreographer may draw on. Another variety of potential energy—spring potential energy—can help us to measure a different kind of physical agency.

Spring potential energy provides a particularly useful way to model the human body: as a coiled spring. When the legs are bent and the person is leaning forward, the muscles in the legs are in a position to unfurl the limbs and launch the person into the air, like a mass on a compressed spring.

Let's begin with a simplified picture of a spring that we can compress or stretch if we do some work. The first component we need to define is the equilibrium position of the uncoiled and uncompressed spring—this is the position it is in when nothing is interfering with it. If we define an x-axis along which it will compress and stretch, the equilibrium position denotes the $x = 0$ position:

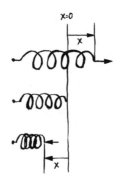

We can therefore use the amount of compression as the variable x moving away from 0. If we are using scientific notation it will be useful to measure x in meters.

What else is important in terms of understanding the energy stored in a spring? For a really stiff spring, a small compression can result in a large amount of stored energy. If the spring compresses without much effort, it will take a larger compression to achieve the same amount of stored energy. The stiffness of the spring is included in our model with the variable k, referred to as the spring constant and with units of N/m, or breaking the newton down into its parts, kg/s^2.

The potential energy associated with a spring (U_{spring}) is:

$$U_{spring} = \frac{1}{2}kx^2 \tag{69}$$

Unlike the situation with gravity, we do not have negative spring potential energy. Regardless of whether we stretch or compress the spring, making x positive or negative, since you square the value the result will be positive. This makes sense because we are working against the spring both when we compress it and when we stretch it. And in both cases we have the potential for motion, whether we let the spring go from a compressed or a stretched position.

Think of the ways your legs can function as springs, storing energy while bent and enabling that stored energy to launch you into motion. What would the spring constant of your legs equal? How would you measure it? Under what conditions would the calculation be meaningful? (At some point, as you bend your legs deeper and deeper you will collapse to the ground and have no stored potential energy for jumping.) You can also try modeling your arms as springs, pushing yourself off from a wall, and considering similar questions. This is how physics is done: we create a mathematical model inspired by observation—in this case our model is the formula for spring potential energy—and test its relationship to movement in the natural world.

Mixing Energies

What happens to the spring potential energy inherent within different dance forms when choreographers blend particular dance styles?

The choreographer Akram Khan synthesizes his training in Kathak, a strain of Indian classical dance, with the British and European contemporary dance techniques that he studied growing up in London. What is actually being combined is not always obvious: Khan will craft a phrase of basic modern dance lunges, for example, but infuse those lunges with Kathak's high-energy vigor. A smooth port de bras or series of whipping arm movements will accompany patterned, rhythmic footwork. Khan is a movement researcher: his body physically integrates the different trainings of seemingly disparate disciplines. His combinations are so seamless that tracing which quality comes from where can be difficult. In refusing to remain constrained within any single dance form, Khan expresses his multiracial identity formed by his Southeast Asian and British roots—an identity forged literally in motion.[20]

Khan offers one version of what the dance scholar Susan Foster calls the "regroove" body. With an eye on dance worldwide, Foster classifies forms by methods of training. Using the terms "ballet body," "industry body," "release body," and "regroove body," she groups dance practices of the twenty-first century and some of their aesthetic effects. Whereas the ballet, industry, and release bodies dominate contemporary dance on stage and screen, the regroove body resists, manifesting "onstage and on YouTube in the riotous proliferation of choreographic gestures that are maintaining the vitality of a myriad of dance traditions...that are now entering the global stage."[21] Foster's effort reveals a dance scholar working scientifically: she deftly links characteristics of movement to history, geography, context of performance, and power structures. Implicit in her assessment is the systematic mobilization of energy that all dance forms carry in their very DNA.

The Energy of Light and the Electromagnetic Spectrum

Just as dance breaks down into individuated forms, various forms of energy in physics can be considered together if we are bold enough to focus in on photons, the particles of light. You are probably familiar with X-rays, radio waves, microwaves, and visible light: these may seem like radically different phenomena, but they are all simply photons traveling through space with different energies. At one end of the spectrum, the high-energy photons include X-rays. While these photons allow us to make images of our internal structure (bones) they are not in a range of energy that allows human eyes to see them. At the low-energy range are radio waves. There are a number of ways to quantify the energies that define these phenomena, but switching from the particle view (photons) to the wave view, we can express the electromagnetic spectrum by the length of the wave, or, correspondingly, by the frequency associated with the light waves.

In order to define the length of a wave, think of waves traveling through the ocean in an area without turbulence, where they roll in regularly repeating patterns. The measurement from one crest to the next crest constitutes the full length of the wave. The frequency, measured in units of Hertz (Hz), can be found if the person doing the calculating stands in one place and counts how many full waves pass by per second. If two full waves pass by in one second the wave has a frequency of 2 Hertz, abbreviated as 2 Hz. A wave oscillating quickly will have a higher frequency than a slowly oscillating wave.

In the electromagnetic spectrum of light, X-rays are high-energy light waves with short wavelengths and high frequencies of approximately 10^{-10} m and 10^{18} Hz, respectively. Radio waves, on the other end of the spectrum, have long wavelengths and low frequencies of approximately 100 m and 1000 Hz, respectively.

Between these two rather extreme cases are the photons, or waves of light, that are visible to our eyes. The wavelength and frequency of visible light are approximately 10^{-6} m and 10^{15} Hz, respectively. The frequency and wavelength of light are also directly connected to what we experience as the color of the light. Higher-energy visible light is toward the blue-purple end of the spectrum, and lower-energy visible light is at the orange-red end of the spectrum.

What happens if a source of light is moving toward us or moving away from us? Just as the sound of a siren in an ambulance driving toward you rises in pitch (and falls in pitch when moving away from you), there is a shift in the frequency of light as the source emitting that light moves toward or away from you. In sound this effect is known as the *doppler effect*. For visible light, there is a *red-shift* when a source of light moves away from us and a *blue-shift* when a source of light moves toward us. A red-shift means that the color of light is shifted to the lower-energy end of the spectrum of light, and a blue-shift means that the color of light is shifted to the higher-energy end of the spectrum. Astronomers see this shift clearly when looking at distant stars and even at other galaxies.

From Marking to Attack

Imagine a dance in which dancers transform into red-shifted or blue-shifted light, depending on their velocity toward or away from the spectator. Of course dancers cannot (yet) dissolve into pure light, aside from the illusions created by costume or lighting design. Nor can choreographers work at the distances or velocities required to physically enact this effect. Dancers do vacillate between lower and higher energies, however, thereby changing a movement's quality and meaning.

Marking is a trade term in dance that means being in the right place and fulfilling the actions with the correct timing but reducing the level of energy applied. Good dancers use this modulation of energy strategically in rehearsal, in order to help them remember and fully understand the movement. By cutting the energy expenditure in half, a dancer can discover more about the spatial dimensions, dynamic, and quality of the choreography. Marking is not only a rehearsal tool: some choreographers build gradations of energy into their final compositions as a way to focus the performers' attention and add subtle texture.

In contrast, moving at full throttle creates a very different effect. In the dance "Cool," for instance, choreographed in 1957 by Jerome Robbins for the musical *West Side Story*, the dancers lunge across the cement and grab at the sky. Coiling and exploding, they are teenagers, members of the gang the Jets, claiming Upper West Side territory as their own. Robbins's movements amplify Leonard Bernstein's musical score and Stephen Sondheim's lyrics: "Got a rocket, in your pocket, keep coolly cool, boy!" "Cool" depicts all-consuming teenage aggression. The gang is a tinderbox, ready to combust. Robbins used energy levels to create character: the Jets can barely contain their own emotional extremes, much less confront the humanity of others.

In 1995, when Robbins restaged excerpts of the musical under the title *West Side Story Suite* for New York City Ballet (in which one of the authors danced the leader of the Jets' girlfriend!), his presence in the rehearsal studio incited yet another version of heightened energy. The dancers knew they were dancing for an artist who was world renowned. Then in his late seventies, he had choreographed over sixty ballets, created and directed numerous Broadway hits, and won a plethora of top awards. He had also gained a reputation for his quick temper during difficult creative processes. The performers' "attack" intensified significantly under Robbins's watchful gaze.

This exercise will help you feel the difference between marking and attacking. Pick one of the movement phrases you crafted in an earlier chapter. Repeat the phrase four times, reducing the energy by the same degree each time. In the first pass across the floor, imagine an irascible choreographer harrying you on. Move with the greatest energy you can muster. Each stride should carry you across an entire metaphoric continent.

Next, try three more passes that modulate the attack, diminishing the energy incrementally each time. On the fourth pass, try marking the phrase. Be sure to hang on to the movement's rhythm and form—change only the level of energy. Making a dance is like laying paint on a canvas; modulating the energy with which movement is executed is the choreographer's equivalent to playing with color, pressure, and line.

Now try a different approach. Break a movement phrase into constituent parts in order to manipulate its energy. How would you go about breaking a dance down? Some choreographers break up a movement phrase into a meticulous series of stills. Others try the opposite, stringing together still images to create a moving phrase, which they then break down again in new ways. Codified steps within certain dance forms are another kind of discrete movement that can be strung together with other movements into choreography. In a choreographic development process, every part of a phrase is ripe for deconstruction and manipulation.

Make a decision about how you will break down your movement phrase into at least four parts. Then vary your energy as you execute the phrase: mark, attack, mark, attack. Just as Akram Khan infused a basic modern dance lunge with the attack of Kathak, you are taking a movement sequence and altering its energetic contents.

Quantized Energy and Bohr's Atom

We have learned from carefully studying physical systems that some quantities, like energy, momentum, and electric charge, are quantized. This quantization means that they cannot occupy a continuous stream of values, but only the values that nature allows. As energy increases, for example, it must jump from one set value to the next. The quantization is largely hidden from us in our everyday lives because it most often applies at extremes, such as at the microscopic scale, or its allowed values are so close to one another that the changes appear continuous.

Quantization is particularly important in physics when a process occurs within a small spatial scale. Investigations into the structure of the atom near the end of the nineteenth century led physicists to image the atom as composed of electrons orbiting a dense, massive nucleus the way planets in our solar system orbit the sun. It was thrilling to think that this orbital phenomenon was repeated on large (solar system) and small (atomic) scales. But this seemingly perfect symmetry breaks down when we look at the details of motion. It turns out that the electrons in an atom behave very differently from planets orbiting a sun. Whereas electrons have specific orbital levels, corresponding to well-defined energies, which they can occupy, the orbits of planets continually decay.

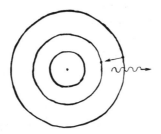

The energy of an electron in an atom, and therefore of the atom itself, is quantized according to these energy levels, as illustrated above in a diagram showing the model of an atom created by Niels Bohr and Ernest Rutherford, Nobel Prize-winning physicists, in 1913. In order for an electron to jump up to a higher energy level the atom needs to absorb energy. For an electron to jump down to a lower level, it has to emit energy, represented in the diagram by the arrow going from an outer circle to an inner circle. This absorption and emission come through photons, particles of light, which are represented here by the outgoing wave. The fact that only certain energy values, and therefore only specific transitions, are allowed means that each atom emits and absorbs well-defined colors (frequencies) of light. These distinct patterns of color give each atom a fingerprint.

Atoms are constructed from protons, neutrons, and electrons and classified into *elements* based on their number of protons. We know the fingerprint of elements like hydrogen, oxygen, and carbon. Remarkably, when we look out to distant light sources in the sky (stars, galaxies) we see these same familiar color fingerprints. We have found these elements in distant constellations and can measure their quantities in the universe.

Energy Images

The gain or loss of energy can cause dancers to "jump orbits" in performance. What do we mean by "jumping orbits" in dance? We are taking a metaphoric leap, here—this isn't the language dancers would use. But the analogy is not far off, for ultimately dancers do jump orbits when they manipulate the level of energy in their bodies, in the sense that they transform not only their movement quality but also their state of being.

How does this work in practice? One strategy is to introduce visual imagery, expressed verbally, for the dancers to focus on that will affect the texture, speed, and quality of their movement. Sometimes called *ideokinesis*—drawn from the Greek words "ideo" meaning idea, and "kinesis" meaning movement—this practice has a long history in the twentieth century, though with different applications. Early theorist-practitioners such as Mabel Todd, Barbara Clark, and Lulu Sweigard used mental imagery to treat issues of skeletal alignment and physical efficiency. Their work entered into the dance world through their teaching and clinical practices.[22] Others began to use the basic principles of ideokinesis to generate choreography.

A particularly striking example of this second type of work can be found in the movement language Gaga, developed by the Israeli choreographer Ohad Naharin. Gaga offers dancers and non-dancers alike a method of physical research that focuses on listening to their body and its sensations. In a Gaga class, the teacher guides the dancers through images that drive and expand their repertoire in terms of speed, texture, presence, and availability for action. Dancers trained in Gaga can have an otherworldly look, with the ability to transform from an average pedestrian to a fantastical creature in an instant.

Gaga classes center around a shared vocabulary of images that Naharin continues to refine, discard, and reinvent over time. The images often work on energy. For example, a Gaga class begins with "floating": an imagined state that is buoyant and weighty at once, as if all movement occurs suspended in water. Other images in Naharin's lexicon include "quaking" and "shaking," "thick" and "soft," "circles and curves," "traveling balls," "the rope of your arms" and "the snake of your spine."[23] Gaga's imagery lifts the practitioners' attention away from forming perfect positions and encourages a nuanced modulation of energy levels. Once the class begins, the instructor and participants move continuously until the class ends. Gaga gives dancers a way to research their bodies, movement capacity, and environment, entirely in motion.

Naharin is not the first choreographer to have created a training method to support his choreographic aesthetic. American concert dance titans such as Martha Graham, Merce Cunningham, George Balanchine, and Katherine Dunham all developed their own techniques. These innovative training methods combine with the choreographic practice to form a kind of ecosystem in which the skills practiced in class feed into the creation of new work, and the creation process informs the content and focus of the class.

But Naharin's method—which he insists is a "movement language" and not a technique—is unique in form and manner of transmission. Rather than steps, Gaga coheres systematically around language, and within that language lie images that change the energy in the dancer's body.

As you watch a dance, pay attention to the quality of movement. Do the dancers move slowly or quickly? Smoothly or jerkily? Do they throw their limbs about or gingerly take new positions? Do they pound downward into the ground or tread lightly? Glide across the floor or strike? Toss themselves off balance or remain vertical? Whirl around, stand upright, curl over, roll, ride momentum, sequence through positions, release and relax, strike and pose, pop and lock, melt, float... Electrons jumping orbits, perhaps?

Floating, condensed, buoyant, constricted, disintegrating: apply these five different states to a movement phrase that you have created. Spend several minutes in each state. Note how significantly the energy and thus the meaning of the movement changes according to the image.

Dark Energy

Consider the unique abilities of the human mind: we can focus our attention on listening to our intelligent bodies and the intricacies of physical research. We can also focus our attention on the energy of the universe and confront one of the fundamental physics mysteries of our time.

The galaxies around us are accelerating away from us and away from each other. Scientists have no idea where this repulsive force could be coming from or what energy is fueling it. It is as if a group of dancers in a room found themselves hurtling toward the walls, away from each other, without their pushing off of the ground. Let's examine this phenomenon, since our evidence for the acceleration is based on ideas that we have already introduced.

Knowing the light fingerprint of each atomic element gives us a tool for understanding the motion of objects in the universe. Remember that light sources moving toward us have their frequencies shifted up, or (in the visible spectrum) toward bluer colors. And light sources moving away from us have their frequencies shifted down, or toward redder colors. When we look at objects in the sky we see the familiar patterns of hydrogen, helium, oxygen, etc., but each of the emitted frequencies is slightly shifted toward the red or blue.

One of the striking observations of the twentieth century was that the farther away celestial objects are from us, the more rapidly they are moving away from us. Physicists have discovered in the details of this red-shifting that the universe is not only expanding, with all objects on large scales moving away from each other, but it is expanding at an accelerating rate. This phenomenon seemingly breaks the rules of energy conservation. For an object to accelerate, we assume that there must be some force acting on it. But the only force that we see acting on these large scales is gravity, and gravity only provides attractive forces. If gravity were the only force at play, objects

113

might still be moving away from each other, but the movement would be slowing down, not speeding up, due to gravity pulling everything in.

If energy is truly conserved, something must be fueling the accelerating expansion. Because scientists do not know the nature of this source of energy, they have simply named it *dark energy*. Researchers are hard at work trying to understand what its source could be and to measure the evolution of the universe's rate of expansion over time.

Real Versus Apparent Energy

Radical ideas about energy are not limited to the world of physics. In the 1960s, the postmodern dance pioneer Yvonne Rainer altered the field of concert dance, in part by focusing on energy. In her seminal dance *Trio A*, which premiered in 1966, she challenged conventional approaches to Western choreographic composition, and in doing so altered the way that audiences are invited to *see*.

The dance begins with a simple action: knees bent, facing stage left, head turned to look over the left shoulder away from the spectators, the dancer begins to move her arms in a metronome-like swing. What unfolds is an idiosyncratic phrase of movement that runs for four and a half minutes, without pause or spectacular effects.

Rainer makes a number of formal innovations in *Trio A* that directly oppose the prevailing dance aesthetics of the day. First, she mixes pedestrian movement with classic dance vocabulary that has been altered into new configurations: a balletic leap with a funny arm grip follows two basic walking steps, for instance, or the hands flap at the ears while the legs and feet assume a classic modern-dance parallel position. Second, she strings these movements together in a *non-repeating sequence*, thwarting the conventional dance composition reliance on theme and variation. Third, she keeps the dancer's gaze averted at every moment, refusing to meet the expectation that a performer will seduce the audience with eye contact.

Rainer binds her choices together with a particularly innovative approach to energy: working against the prevailing conventions of Western classical and modern dance aesthetics that insist on the "rise and fall" of dramatic timing, she deliberately evens out the distribution of energy in the dance. The dancer seems simply to transition from one movement to another, as if running an errand or going about her day.

In *Trio A*, Rainer expresses a theoretical statement choreographically. She illuminates her polemics in a short essay (with a long title): "A Quasi Survey of Some 'Minimalist' Tendencies in the Quantitatively Minimal Dance Activity Amidst the Plethora, or an Analysis of *Trio A*." At stake is the entirety of Western classical dance and its exhibitionist display. She writes: "Like a romantic, overblown plot, this kind of display—with its emphasis on nuance and skilled accomplishment; its accessibility to comparison and interpretation; its involvement with connoisseurship; its introversion, narcissism, and self-congratulatoriness—has finally in this decade exhausted itself, closed back on itself, and perpetuates itself solely by consuming its own tail."[24]

To challenge these "overblown" aesthetics, Rainer redefines the distribution of energy within a movement phrase. She is most concerned with "real" versus "apparent" energy. While classical ballet and modern dance foregrounded apparent energy—the illusion that a dancer expended no effort at all to accomplish the most difficult feats—Rainer was far more interested in trying to represent actual energy: a dancer squatting, rolling, rising, jumping in the time that it actually took to squat, roll, rise, and jump. In lieu of a dynamic that theatrically rises and falls with each step, Rainer substituted a purposeful, workmanlike *doing*. The dancer's energy expenditure became an undeniable fact, rather than an illusion or artifice.

Rainer's *Trio A* asks viewers to rethink their expectations of what dance is and should be. We can think of her move as a dance parallel to Albert Einstein's $E = mc^2$ proposition, because both alter how we see our world. The four-and-a-half-minute treatise that is *Trio A* has influenced artists across dance, visual art, and film.

$E = mc^2$

We have seen throughout this chapter that energy fuels and shapes motion. But energy is not limited to shifting from one form to another, such as kinetic to potential, or vice versa. In physics we also can have energy sloshing back and forth with mass, governed by one of the few physics equations that seems to have infiltrated pop culture, Einstein's famous

$$E = mc^2 \tag{70}$$

This equation gives us an equivalence between energy E and the mass m multiplied by the speed of light c squared. This equation is particularly useful in research in which fundamental particles are accelerated to near the speed of light in underground tunnels and smashed into each other to help physicists explore the building blocks of nature and the forces with which they interact. The kinetic energy from the colliding particles is available for nature to play with. Where there was energy, there can be a new, massive particle created. The amount of kinetic energy available puts an upper boundary on the mass of the particle that can be created. If physicists want to look for even more massive particles than they have already discovered, they will have to make the colliding particles travel faster, therefore creating more kinetic energy, before smashing them together.

Particles are created in these collisions from available energy, with the probabilities of their creation dictated by the laws of nature. Very few of these particles are stable: they decay, leaving different particles with their own masses and kinetic energies behind. Acceleration, collision, the birth of new particles and their subsequent decay all happen under the watchful and quantifying eye of the experimenters who are trying to understand some of the mysteries of the universe. This is the rhythm of particle physics. We might even call it a dance.

8. Space

Physicists at the Relativistic Heavy Ion Collider at Brookhaven National Laboratory in Upton, New York, put gold atoms into powerful electric fields that rip away their electrons. The remaining gold nuclei are accelerated to near the speed of light and steered into each other. The resulting collisions produce temperatures thousands of times hotter than the temperature of the sun. These collisions allow scientists to study a primordial soup of matter believed to have existed fractions of a second after the Big Bang.

To the observers in the lab, the accelerated gold nuclei at Brookhaven are flattened like vertical pancakes. Relativistic objects—objects traveling at speeds that approach the speed of light—do not maintain their shapes as viewed from the perspective of observers watching them fly by. They are flattened in the direction of their travel due to the effects of special relativity. Physicists have come to understand that an object's spatial measurements change along with its relative speed. The study of special relativity has altered scientists' perception of the nature of space from something they thought of as inherently absolute to something that is malleable, changing according to the perspective of the observer.

Radical experiments with space also fill dance history. Whether through abstract compositional play or attempts to literally render ideas drawn from astrophysics, artists have pushed the human form in pursuit of far vaster realities. The scientific zeitgeist frequently captures artists' imaginations: as early as 1932, the dancer-choreographer Ruth Page and the sculptor Isamu Noguchi drew on advances in modern physics to create *The Expanding Universe*. Their ambition was to depict nothing less than the accelerating expansion of the cosmos. Noguchi designed for Page a wearable sculpture—a shimmering blue sack of a dress that covered most of her body, leaving her head free. To represent the volatile, active nature of spacetime, Page moved within the supple fabric so that it folded and rippled.[25] Her body could hardly dissolve into the vacuum of outer space: her classic modern-dance movement vocabulary gave away her historical place and time. Still, Page and Noguchi's attempt to humanize the infinite universe is a notable example of artists drawing on science to rethink representations of space.

In this chapter, we examine the role that space plays in the work of artists and physicists, beginning with the smallest possible scales of matter and ending with the largest. We start with the micro-attentions of dancers and

116

an investigation of subatomic physics, move through a Native American circle dance and dance notation, and conclude with the concept of length-contraction in Einstein's theory of special relativity and artistic challenges to the proscenium stage.

As we progress through the chapter, we will also be moving through different ways of knowing space, from experiential and aesthetic to conceptual and mathematical. Dancers use movement to explore both inward and outward frontiers—excavating the psyche as much as the environment in which the dancing occurs. Science, meanwhile, has relied upon ever more sophisticated technologies and mathematics to uncover fundamental spatial principles of nature, which they could not access through experiential observation.

The vastly disparate means by which humankind has gotten to know space represents a major difference between physics and dance. We cannot have a sensorial encounter with particles, nor can we be present in outer space to witness physically the collision of two black holes. We need powerful imaginations for this aspect of our study. Indeed, the human imagination is one major bridge linking quantitative, aesthetic, and embodied ways of knowing.

Dancing Cells

It may seem obvious that choreographers must contend with the limitations of human anatomy. They must figure out where their dancers' arms, legs, and heads will go, and with what rhythm, coordination, and nuance they will move. Less obvious, however, are the motivations that drive movement. We cannot perceive with our human senses the cells and atoms that make up our being, but some choreographers have in fact drawn on this microscopic scale to inspire their art. The experimental American choreographer Deborah Hay offers perhaps the most extreme example.

Hay playfully interrogates and responds to her *cellular body* to incite her movement. She does so using specific questions to launch her into motion: "What if," she asks, "every cell in my body had the potential to perceive time passing?"[26] She recognizes the absurdity of her prompts; the questions are impossible to answer. It is the effort to question through her body, and the movements that result, that matters to her art.

Hay is not researching cells or cellular space but rather exploring the effects of language upon the dancer's bodily consciousness. Understanding that verbal imagery affects a dancer's kinesthetic imagination, she poses questions that are both purposefully worded and open ended. A question such as "What if every cell in your body at once has the potential to perceive your movement as your music?" could generate rhythmic play, as if the dancer were listening to an unheard melody.[27] Other prompts are motivated by more pragmatic concerns. When Hay wanted to challenge her hardwired instinct to face the audience—the typical onstage orientation—she questioned her assumptions at the level of her cells: "What if every cell in my body at once has the potential to choose to surrender the pattern of facing a single direction?"[28] Over her five decades of making work, Hay has adapted her imagery according to advances in biological science. In the 1970s, she worked with the image of 5 million cells; in the 1990s, 800 billion. Today, she imagines "more than a zillion."[29] The number of cells matters less than the attention that Hay's dancing with cells elicits.

Hay's questions inject her movements with extraordinary focus. When she dances, she appears to be intensely tuned, probing, and whimsical, motivated by dramatically different means from those informing the typical steps and phrases of conventional Western dance. By tuning in to the infinitesimal, she creates a delicate connection between body and mind. A viewer would never guess her cellular prompts—her impulses remain mysterious. Hay ultimately choreographs consciousness.

Hay's engagement with human biology proposes an alternative to linear reason. In throwing off prescribed patterns of thought, she implicitly argues for a different approach to thinking itself. In her art, objectivity becomes a biofeedback loop between the dancer's own body and psyche.

A Glimpse into the Microscopic World

Hay's artistic practice, while not a scientific investigation, does share similarities with the scientific method: a hypothesis is stated and tested, and conclusions are drawn. Scientists have turned this process toward the task of discovering the content of the universe.

One of the biggest surprises of eighteenth-century physics came from an attempt to determine how matter is structured. Physicist J. J. Thomson had devised a "plum pudding" model, in which he imagined a positively charged substance, like a pudding, with negatively charged electrons floating around like plums floating in pudding. If the positive pudding canceled the negative plums, what remained was the neutral, or uncharged, matter that he knew existed in nature.

Ernest Rutherford conducted a famous experiment to test this idea. His team directed a beam of positively charged bits of matter (known as alpha particles) at a sheet of gold foil and noted how the particles interacted with the foil.

To understand this experiment, it helps to put yourself in the position of one of these alpha particles. Imagine that you are in a large, dark room. You attempt to learn about your surroundings by running from one side of the room to the other. If the room is empty, with no obstructions, you could go back and forth without changing your path. If you ran past the blast of an air conditioner, it would not have much impact on your path. If you ran into a table, you might be deflected into any of a number of different directions. If you instead ran into, say, a cluster of punching bags hanging from the ceiling, you would only be able to pass through them if you started out with enough momentum to break through the barrier. And if the barrier were a trampoline set on its side, you would be bounced back in the direction you came from. (Here we note that it is not a great idea to run around in a darkened space without knowing what you might encounter.)

Even if it were safe to run through this dark room, you are far too big to serve as more than an analogy when it comes to probing the nature of matter on the atomic scale. That is where an alpha particle has you beat. In our analogy with Rutherford's experiment, the alpha particle is you, and the foil is the room you are trying to explore. Scientists found that most of the alpha particles they sent into the gold foil passed through with very little deflection, but approximately one in twenty thousand bounced back, way off course. As Rutherford later wrote, "It was quite the most incredible event that has ever happened to me in my life. It was almost as incredible as if you fired a 15-inch shell at a piece of tissue paper and it came back and hit you."[30]

The new model of the atom that emerged from these experiments concentrated the bulk of the mass in a very small volume at the center—the nucleus—rather than spread out evenly over space. The tiny fraction of alpha particles that bounced off in odd directions must have hit the small, massive nuclei as they interacted with the foil, but the majority of the alpha particles were fired right through empty space.

Thomson and Rutherford's experiments provided a blueprint for the development of nuclear and particle physics. By the twentieth century, physicists had developed sophisticated research tools, including particle accelerators and detectors, that allowed them to probe smaller and smaller spatial dimensions. Even in a vacuum from which all of the air has been sucked out, activity takes place within these smallest spatial dimensions. In order to model nature on that scale, we must move from Newtonian mechanics to quantum mechanics, where things become unpredictable. In a process physicists describe as "quantum mechanical fluctuations," particles can even be created for a brief period of time by borrowing energy from the vacuum. "Empty space" is not so empty.

The Geography of the Body

It can be a shock to realize that Newton's laws and the ordered universe that they predict break down at subatomic scales. No parallel, scale-induced shock can be found in the discipline of dance, for even when imagining their cellular bodies, dancers must work firmly within macro-world laws of motion. Within those laws, however, the human body and its seemingly infinite potential for motion can transcend space—evoking multiple spaces at once, or a kind of *polyspace*, as in the work of the choreographer Garth Fagan.

Working on a vastly different scale from Deborah Hay, Fagan imagines the human body as an intricately linked map of diasporas. *Diaspora* describes movement and displacement—in particular, patterns of migration, which may be voluntary or forced. It is frequently used to refer to the African Diaspora, marked by the tragic dislocation caused by the transatlantic slave trade that began in the sixteenth century and spread Africanist aesthetics worldwide. Born in Jamaica and having immigrated to the United States in his late teens, Fagan has lived through a personal diaspora. He studied with the modern-dance masters of the mid-twentieth century, including Martha Graham, Katherine Dunham, and Alvin Ailey, and he also counts Caribbean and West African dance as major influences. Fagan charts his travels choreographically by blending movements drawn from his own personal training in American modern dance, Afro-Caribbean dance, and classical ballet.

Fagan's dances do not just flow from one style to the next, however: he cuts up and remixes styles in a single dance. A supple, Africanist torso might accompany a delicate balance in relevé, creating the image of reeds waving in the breeze; or a rhythmic pulse in the pelvis might pair up with a deep, turned-out plié.

When taking in Fagan's choreography, it is tempting to identify movements according to their geographical origins, noting that the circles of the pelvis come from here, or the rise in the shoulder blade comes from there. But such an exercise overlooks his main point, which is an argument about the way diaspora works. The transmission of cultures across space is not abstract: people embody diaspora in the shoulder rolls, hip thrusts, and balances of a dance. This embodiment allows Fagan to move his Afro-Caribbean culture beyond the geographic limits of the Caribbean.

But Fagan is also offering an argument specifically about the power of the artist. For Fagan does not simply pick up influences as he travels, as if he were a blank slate upon which culture is written. He bends culture to his purposes. One of his earliest pieces, *From Before* (1978), captures his mastery over his movement material. Fagan uses the values of minimalism, an aesthetic movement of the mid-twentieth century that spread across visual art and dance, to magnify the pure formal power of his Afro-Caribbean heritage. He distills Caribbean dance down to its polyrhythms, isolation of anatomy, and fluid spine, while drawing on the precision of ballet to support his dancers' feats of strength. *From Before* valorizes Caribbean knowledge, even as Fagan also claims minimalism, often stereotypically associated with white artists, as his own. When his histories, places, and cultures collide in the bodies of his dancers, Fagan's art turns "space" into "polyspace."

Dancers in the Universe

In Garth Fagan's art, any given movement evokes vast geographies. Opening the scope of our inquiry into space and dance even further, how do dance artists situate their dance within the universe?

There are as many ways to answer this question as there are dancers in the world. We are moving our inquiry outward now, away from the dancer's internal, physical strategies and techniques toward real and imagined spatial formations that dancers inhabit as they move. Two strikingly different examples will illustrate this point: the Native American Ghost Dance and Rudolf von Laban's kinesphere.

The Ghost Dance is an empowering dance and Native American religion that had become a pan-Indian movement by the end of the nineteenth century. At the heart of the religion lies the practice of dancing in a circle. The dance first emerged in the 1870s through a leader of the Paiute of western Nevada, but the practice had been documented earlier in Native American history, including previous circle dances on record from the early 1800s: the Prophet Dance and the Great Basin Round Dance.[31] Ghost Dancers later adapted these rituals and circular formations to meet new needs and new doctrines. After emerging in the 1870s, the Ghost Dance disappeared and then resurfaced in the late 1880s, revived by a Paiute prophet named Wovoka. Identified in the community as a messiah, Wovoka believed he had been called by God to spread a doctrine of peace and a vision for a bright future to be enacted through a ritual dance.[32]

Native eyewitnesses from the 1870s describe the dance as a series of concentric circles, with every other circle rotating in the opposite direction, sometimes containing as many as ten circles.[33] In an 1890s gathering, up to three thousand men, women, and children might participate, thus requiring a vast site in which to perform the ceremony: ideally a flat terrain, cleared of trees and brush, with easy access but also secluded, so as to shield the dance from non-native eyes.[34] Ghost Dancing incorporated rainmaking songs; white body paint; "eagle, crow, sage hen and magpie feathers"; and a magical "bulletproof" shirt that was painted with, among other signs, Numic concentric circles.[35]

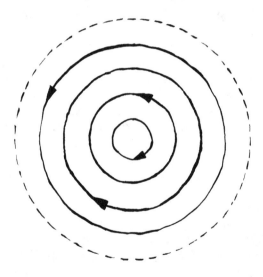

Historians widely agree that the Ghost Dance emerged in reaction to the trauma of forced migration and the assault on indigenous culture caused by American westward expansion.[36] But U.S. government officials at the time poorly understood the dance and saw it as brewing aggression toward the white populations who were moving into Native lands. In 1890, the U.S.

Army attacked and killed over two hundred Lakota Sioux in an event now called the Wounded Knee Massacre—a tragedy largely instigated because of the misperceived threat of the Ghost Dance.

In contrast to the forced containment of Native Americans, the Ghost Dance knew no boundaries: it spread geographically across the continent, and envisioned a porous boundary between the worlds of the present and the afterlife. Forging a cosmic connection, Ghost Dancers sought links to their ancestors, channeling the Numic vision of *Puha*, or power, a force that Native Americans understood not just as a mechanism but also as the essence of the cosmos.[37] The circle dance connected tribal members to the all-pervasive powers of the universe.

Thirty years later, on another continent, the German-based choreographer and theorist Rudolf von Laban imagined universal power very differently. Laban's choreography is less remembered today than is his major contribution—a system for writing down dances known as Labanotation. His movement analyses originated out of his practices of dancing, choreographing, and teaching while also drawing on his knowledge of Euclidean geometry, mathematics, physics, and human anatomy. He aspired to make his notation universal, capable of transcribing any and all human movement in the world.[38]

Put yourself in the shoes of a choreographer-theorist of the 1920s, and imagine trying to develop a system for extracting and documenting data from human movement. You have no digital technologies to assist you. How would you distill down the basic characteristics of all dance, given the many cultural variations and idiosyncrasies that we have considered? Any good notation would need to account for the shape, quality, and duration of a movement. You would also need to record spatial relationships between dancers and among the dancers and their environment.

In the centuries before Laban, many people tried to record dance in print. The majority of these systems were designed to document specific dances or genres, and could not fully handle the discrepancies in quality, rhythm, and dynamic that differentiate one movement practice from another.[39] The dance maker and notator Raoul Auger Feuillet published a system at the turn of the eighteenth century, for instance—building on an earlier effort by the French ballet master Pierre Beauchamp—which mapped the patterns of the feet on the floor, but failed to capture the nuances of the dancer's arms and torso.[40] Laban's method tracks the upper body as well as the legs, especially shape making (which he refers to as "traceforms"), and uses symbols to record effort, energy, direction, and spatial patterning, among other aspects of human motion.

Laban had to contend with one major problem: all human movement has to occur in a specific location. He concluded that space and time were inextricably linked. Space, he wrote, "is a hidden feature of movement and movement is a visible aspect of space."[41] Laban was referring exclusively to human movement, but he might also have been thinking about the cosmos:

from human action to planetary ellipses to post–Big Bang expansion, the universe comes into view through motion.

Laban linked human movement to the cosmos through his kinesphere, a geometric organization—diagrammed as a cube—that surrounds an individual's personal space. The kinesphere emanates outward from the body according to basic orientations in space—length, breadth, and depth. Each dimension runs in two directions: up and down, left and right, forward and backward. Diagonal lines also run through the cube. All points converge in the body's center of mass, which lies at the center of the cube.[42] The kinesphere frames human movement by positing a grid in which actions occur.

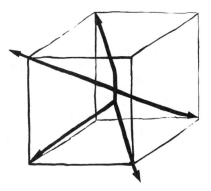

Even as Laban imagined this cube enclosing the human body, he saw a link between the kinesphere and the universe. He wrote, "Innumerable directions emanate from the centre of our body and its kinesphere into infinite space."[43] Wherever we go, he suggested, our kinesphere follows us, and with it comes an awareness of the vast universe beyond our own. In connecting the human form to a far more expansive reality, Laban implied that we can, through our own small mortal being, tap into the infinite.

Laban's totalizing vision for movement analysis developed in the context of a dramatically shifting political landscape. He worked first under Germany's Weimar Republic of the late 1920s and 1930s, and then under the rising Nazi regime. Historians debate Laban's complicity in the Nazi's nationalist project, and the extent to which high-profile artists in Germany might have felt pressured to comply with the state remains in question.[44] Laban eventually fled Germany for England, where he completed his writings on movement analysis and started an important center for dance.

Labanotation is still used today to record dances. Even as video has become readily accessible for performance documentation, his notation extracts other information that might not be discernible on camera. Subsequent choreographers have also picked up and manipulated the idea of the kinesphere by moving its center to other parts of the body, reducing the cube in size, and even surrounding the dancer's body with multiple kine-spheres at once. Laban's construction of space has become a creative tool that artists have built upon, questioned, and challenged, revealing that his human-centered universe, informed by Western geometry, is anything but nature-made.

Laban and the Ghost Dancers propose different ways of knowing space. Both suggest that the human body in motion can yield insights into the cosmos beyond planet earth. (Remember that Newton, too, extrapolated from his observable experience on earth outward to the moon, when he developed his Law of Universal Gravitation.)[45] But their visions of space also differ in important ways. Laban imagines a universe, and the power to universalize, centered on the individual: one person in one cube. Ghost Dancers structure space collectively. The circular shapes in which they dance must be built together, with participation from the living and the dead.

Ultimately, Laban fell short of his goal to create a universal movement notation that could record all cultural forms. Certain details will always escape capture. Labanotation could document the spatial formations and steps of the Ghost Dance. But the spatial imagination bound up in those patterns—with their ties to the afterlife, their profound longing for sanctuary, and their claim on the future—would remain out of reach. The concentric circles of the Ghost Dance remind us that where there is space, there are struggles for power over that space. Sometimes the resistance takes the form of a dance.

Room Writing

We can focus on internal or external space in dance, but in the end inside and outside are mutually informative. A viewer can read in your movements what you are thinking when you move. And the space in which you dance affects how you move and thus how you think. How might we investigate these ideas in a movement exercise?

You are going to *scan* the geometry of the room in motion. What does this mean? The idea is borrowed from the choreographer William Forsythe's improvisation technologies. Forsythe calls the exercise "room writing," and it entails searching for shapes in the room to mark with your body, in motion.[46] Any shape is fair game: you might see and delineate a circle with your left elbow, or a cube with your right knee. You could attempt to mark the crosshatches of a heating ventilation system, like the one in our dance studio, with your spine. The aim is to see and respond to your environment through your body in motion. You might mark discrete points in a pointillistic fashion, or perhaps sweep a shape into view and then allow it to dissolve. Whatever you choose to do, keep moving.

The goal is to realize the visual information in movement form, however and whatever that means to you. It is important that you stay true to your task. Simply mark geometries in the room. This requires attentiveness, observation, and the creative ability to translate the shapes that you discern into movement form. Do not worry about the viewer's being able to tell exactly what shapes you are translating. The most important thing is that you are thinking, processing, and imagining in motion. Only later will you compose—which is to say, figure out how to do something with the movement ideas you are activating.

After working for a number of minutes on this idea, rest a moment. The second phase of this exercise has to do with exploring volume. Volume is a choreographic idea borrowed from geometry, in which it refers to the three-dimensional space within a contained frame. When dancers reduce the volume of movement, they do not lower the intensity of execution; rather, they shrink the space in which the choreography is performed. Magnifying the volume means expanding the amount of space that the movements must attempt to fill.

Conjure your own personal kinesphere, which should be slightly larger than your entire figure, and imagine and center yourself within this cube. Work on scanning the geometry of the room within this kinesphere, which extends from the tip of your fingertips to your toes. You might indicate the points on the cube with your arms, legs, head, elbows, knees, toes, chest, ears. There are many angles at which to direct your movements. Work at this task for a while. It is the effort that matters, not achieving perfection.

Pause a moment and recenter yourself within your kinesphere. (Would it not be comforting if we could all simply step inside the security of our own personal imaginary cube?) Only now, imagine your kinesphere shrinking in around you—this new region frames your shoulders down to your knees, and all of your actions must take place inside these new dimensions. Try scanning the environment again at this lower volume.

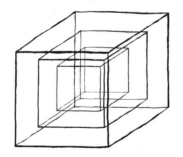

We have just changed the scale from reaching out into Laban's infinity, to a smaller, more interior space. Moving within this smaller cube requires much less reach, and that gives your movements a more intimate feel. Setting the coordinates of the volume is ultimately up to the creator. They are arbitrarily assigned, just as they are in physics, to achieve different movement qualities and meanings.

Postulates of Special Relativity

If moving their arms in the kinesphere somehow taps people into the pulse of the universe, what are they sensing "out there" when they move? The cosmos is infinitely vaster, and space and time more warped and relative, than humans can fathom through experiential evidence alone. Einstein needed mathematics to solve the problem of relativity, but he never left the sensation of movement far behind. He used his kinesthetic imagination, just as a choreographer does, to develop thought experiments that involved both real-world and cosmic motion. These imaginary scenarios played an important role in his theorizing.

Einstein's special relativity brought physicists the notion of a universe-wide speed limit and a fundamental link between time and space. He also shattered the current ideas of absolute time and space. You, too, can reach the conclusions that Einstein reached, if you can understand a few of his postulates and embark on an exploration using algebra.

In order to understand Einstein's postulates, you must first understand the concept of an *inertial reference frame*. A reference frame is a coordinate system that is centered on something. You have just been introduced to such a frame, Laban's kinesphere, centered on your pelvis, which you carry around with you as you move through space. Everything around you is in relation to this cube, the position of which you control with your motion.

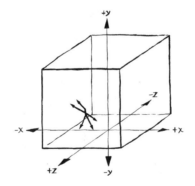

You could also consider a frame that is independent of your motion—say, a dance studio. You would measure all movement within the studio with respect to the room. You could set the positive y-axis as pointing directly toward the ceiling, with the floor defining $y = 0$. The floor itself makes up the x-z plane. If you are standing still in the room, you have zero velocity with respect to the reference frame defined by the studio. As you move, you would define your velocity and acceleration with respect to the stationary room. Imagine moving through a room that you have delineated in your mind with this fixed grid. You can move along axis lines, traveling parallel to the wall, or cut across them diagonally, with your position at any point in time marked by the intersection of lines from your imagined grid.

In this next mental exercise you will imagine a number of other reference frames. Zoom out from the studio, put on your spacesuit, and imagine that you are standing on the moon looking at the dance studio on earth. Redefine your reference frame so that the center is now located with you on the moon. The moon is, and therefore you are, locked in orbit with the earth. You see the earth, the studio, and the dancers inside it slowly rotating underneath you. Your view of the earth, and thus the studio, changes as the day progresses, and you could measure the studio's velocity in your reference frame, which is centered on the moon.

You could also fix your reference frame with the sun at the center. The dance studio on the earth, the earth itself, and the moon are all in motion with respect to the frame of the sun.

You have a lot of flexibility in terms of defining your reference frame, so choose the one that makes your calculations simplest. For example, if you are analyzing the motion of a dancer in a studio, establishing the room as your frame of reference would make sense. If you are analyzing the motion of planets around the sun, or a dance studio on earth from the perspective of the sun, then centering your frame on the sun makes sense.

An *inertial reference frame* is a nonaccelerating reference frame. You are an excellent judge of whether your reference frame is accelerating or not. Have you ever been sitting down, minding your own business, when all of a sudden—with no one pushing you—you are thrown in one direction or another? It happens in buses, trains, trams, airplanes, and cars, any time the vehicle you are in swerves, speeds up, or slows down. These are all moments of acceleration, and you are good at detecting them because the normal laws of physics seem to break down. For example, your water bottle on the floor of the car goes skidding to the front of the car when you slam on the brakes, or your bag slides off your seat when the bus makes a turn.

For another exercise, picture dancing within a studio that has been lifted onto the bed of a very large train and is moving quickly down a curving track over a hilly landscape. You would sense that something different was going on even if your studio was completely soundproofed and had no windows: you would one moment be standing in one position, and the next moment be thrown in one direction or the other. In an inertial reference frame, such as the dance studio planted with a good foundation on the surface of the earth, you would not suddenly lurch to one side or another (unless there's an external force acting upon you within that reference frame—such as another dancer giving you a shove, or something like an earthquake that provides an acceleration of the studio). If the train carrying your studio starts to travel at a constant velocity along a straight and flat portion of the track, your frame would be nonaccelerating as long as that constant velocity was maintained. You would be in an inertial reference frame. You would have no idea, within the studio, that you were in motion with respect to the earth. As soon as the

127

train took a turn, you would no longer be within an inertial reference frame. You would be able to sense this by seemingly unexplained forces acting on your body resulting from the frame's acceleration.

With this working definition of an inertial reference frame, you are ready for the postulates of special relativity. (Postulates are the assumptions upon which a theory is built.) In this case, we will ask you to trust Einstein that these postulates are a valid starting point while you work through their framework-shattering implications:

Postulate 1: The laws of physics are the same in all inertial reference frames.

Postulate 2: The speed of light in a vacuum is the same in all reference frames.

These postulates have held up against a torrent of tests. So far they seem to be correct. They may seem straightforward at first glance, but their implications are astounding.

In order to understand the first postulate, consider the feeling of stillness. Take a moment now, right where you are, reading this book, and pause. Try to find stillness. Feel that all forces acting on you are equally balanced and that your body is not accelerating. (This will be difficult if you are not currently in an inertial reference frame.)

Now that you have established the feeling of stillness, remember that you are, in fact, on a planet that is orbiting the sun, in a solar system that is moving through the galaxy, in a galaxy that moves through the universe. How fast are you moving? You cannot actually answer that question without knowing the answer to "with respect to what?" How fast you are going with respect to the street will have a very different value from how fast you are going with respect to the center of the black hole that is theorized to be at the center of our galaxy, the Milky Way. The universe is not a place with well-defined boundaries or a meaningful center, and there is no definite spot of stillness in it. If the laws of physics are the same in every inertial reference frame, there is no preferential reference frame. The first postulate of special relativity tells us that there is no special reference frame that defines true stillness. Everything is, well, relative!

The first postulate of special relativity carries good news and bad news for physicists. The good news is that the work they do to understand the laws of nature in one reference frame should apply to all other reference frames, giving it an expansive reach. But at the same time, they are unmoored: there is no center of the universe or reference frame that physicists can view as privileged.

The second postulate of special relativity requires more effort to understand. It states that the speed of light remains constant no matter where you are in the universe: someone on earth and someone sitting on a comet passing by will come up with the same measurement of the speed for each photon, or particle of light. To explain the implications of this assumption, we have set up a performance.

Special Relativity: Length Contraction

The featured performer is in the middle of a dark dance studio, low to the floor with a flashlight in hand. There is a mirror on the ceiling. The performer will turn on the flashlight. A second performer, with a stopwatch, will measure the time it takes for the light from the flashlight to travel up to the ceiling, bounce off the mirror, and return to the point where it was released. The audience is sitting on the edge of the studio, each person with a stopwatch to confirm the measured time. This scenario requires a few leaps of the imagination: it is necessary to assume that the stopwatches are extremely precise and the performers' and audience members' reaction times perfect, to clarify the implications of the second postulate in this performance art piece.

The light must travel the distance from the flashlight tip to the mirror and back. If we label this distance from the flashlight to the ceiling d, then the total distance traveled will be $2d$, representing the round trip. We can label the total time that the performer's partner measures as t_p (where t stands for the time in units of seconds and the subscript p indicates that this is the reference frame of the performer). We can calculate the velocity of the photons because we know how far they went and how long it took them to make the trip. Remember that velocity is equal to the distance traveled divided by the time it takes to travel that distance:

$$velocity = \frac{distance}{time} \tag{71}$$

We shall now write down the formula for the speed of light, denoting it as v_1 to correspond to our first performance scenario:

$$v_1 = \frac{2d}{t_p} \tag{72}$$

Thinking like choreographers, now we add another layer of complexity to our performance. If we return the studio to the flatbed of the train, traveling with constant velocity, this performance—and our calculations—become substantially more interesting. The audience will now be located outside of the studio, watching it, and the performance, zip by on the train.

For the first performer and the partner, nothing has changed, assuming the train does not bounce, turn, speed up or slow down. The dancers are in the inertial reference frame of a nonaccelerating and therefore constant-velocity train carrying their studio. The audience is also in an inertial reference frame on the side of the tracks, but the spectators see the light take a different path. Instead of traveling straight up and down, as it does from the performer's perspective, they see the light released at one point in their frame and then hit the mirror some distance down the track. The train has traveled even farther by the time they see the light return to the flashlight. They therefore see a path that can be described as the hypotenuse of a triangle, as shown in the diagrams that follow:

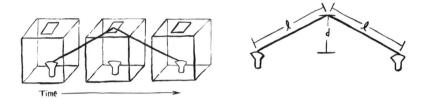

Time ───────────────►

If we label the path of the light on its way up as l, it is clear from this diagram that l is greater than d. The total distance that the audience sees the light travel is now $2l$. With their stopwatches, the spectators measure a time t (where t stands for time in units of seconds). We can, again, calculate the velocity that the light is traveling:

$$v_2 = \frac{2l}{t} \tag{73}$$

Note that the faster the train goes, the larger the value for $2l$ will become, because in the amount of time it takes for the light to travel up and bounce back the train will have gone farther down the track. As the train approaches the speed of light there will be a more and more significant difference in the lengths that the light travels in the two different reference frames.

The second postulate of special relativity holds that the speed of light is measured to have the same value in every reference frame. This means that $v_1 = v_2$ because both of these velocities are measures of the speed of light. But this cannot be possible, because the light travels different lengths according to the different observers. The staggering way out of this impasse is to concede that time is not absolute. The various observers, each with a perfectly working stopwatch, are actually measuring different times. This well-established effect is known as *time dilation*, and we will focus on it in the next chapter.

Another strange phenomenon happens when you watch the performers zip by in the dance studio. If their speed is close to the speed of light, you see them flattened like pancakes in the direction of travel, and the dance studio becomes shorter. This effect is known as *length contraction*. Here comes your algebra.

First define the variable gamma (γ), which takes into account how fast the two inertial reference frames are moving with respect to each other (such as the speed of the train versus the audience on the side of the track) and the speed of light. The variable γ, used throughout calculations in special relativity, is defined as

$$\gamma = \frac{1}{\sqrt{1 - \frac{v^2}{c^2}}} \tag{74}$$

130

where v is the relative velocity between the two frames of reference and c is the speed of light. In our example, v is the velocity of the train, because that is the velocity of the reference frame of the studio with respect to the reference frame of the ground. We will work with velocities in units of meters per second. In those units, the speed of light is:

$$c = 2.99 \times 10^8 \text{m/s} \tag{75}$$

What values can γ take? If the train is not moving at all with respect to the audience on the ground, the velocity of the reference frame of the studio will be 0 with respect to the audience's frame. When $v = 0$, we can see that v^2/c^2 equals 0. That means that γ equals 1. If, however, the velocity of the train is close to the speed of light, we can see that the value of v^2/c^2 can grow. Since the train and the dance studio it is carrying have mass, they can never get up to the speed of light, so the velocity v will always be less than c. This is important because it means that v^2/c^2 can never reach the value of 1. Thus, in the denominator of the formula for γ, we can see that we will never have to deal with a negative number within our square root. (If we had a negative number in our square root we would need to worry about imaginary numbers, and we have enough to worry about.) The variable γ will have a value equal to or greater than 1.

Let's now consider how γ can be used to calculate lengths and quantify length contraction. There are two lengths that we need to keep track of in this problem:

L_P: The length of the dance studio as measured by those in the studio. This will be known as the *proper length* of the dance studio. A length is "proper" if it is measured by someone who is in the same reference frame as the object being measured, so the object is at rest with respect to the measurer.

L: The length of the dance studio as measured by the audience, the observers in another reference frame. We can simply refer to this length as L.

The values of L and L_p are related to each other through the variable γ:

$$L = \frac{L_p}{\gamma} \tag{76}$$

The length L that an observer measures will always be less than the object's proper length L_P, if the observer is moving with respect to the object, because γ will be always be greater than 1. Remarkably, even simply walking past someone on the street will affect the observer's measurement of that person's width along the direction the observer walks—but the effects are far too minuscule to see. The difference between L and L_P becomes significant only when relative velocities approach the speed of light.

Returning to our dance performance on the moving train, if the dance studio is 20 m long (about 60 ft) and the velocity of the train is just 1% of the speed of light, then $v = 0.01c$. (This means that v is equal to 2.99×10^6 m/s, already a whopping 6.7 million miles per hour. Note that it's far simpler in the calculation if we report v as a fraction of c instead of plugging in the actual values for v and c, which can create significant digit problems on a standard calculator.) The γ value would be

$$\gamma = \frac{1}{\sqrt{1 - \frac{v^2}{c^2}}} = \frac{1}{\sqrt{1 - \frac{(0.01c)^2}{c^2}}} = \frac{1}{\sqrt{1 - 0.0001}} = \frac{1}{\sqrt{0.9999}} = \frac{1}{0.99995} = 1.0001 \tag{77}$$

That means that the audience would measure the studio as approximately 19.998 m long instead of 20 m, according to the following calculation:

$$L = \frac{L_p}{\gamma} = \frac{20.0 \text{ m}}{1.0001} = 19.998 \text{ m} \tag{78}$$

This is extremely close to the proper length of 20 m. If, however, the train is moving at 50% of the speed of light ($v = 0.5c$, or 1.495×10^8 m/s) then we see a much larger impact. First we calculate γ for that relative velocity between the two frames:

$$\gamma = \frac{1}{\sqrt{1 - \frac{v^2}{c^2}}} = \frac{1}{\sqrt{1 - \frac{(0.50c)^2}{c^2}}} = \frac{1}{\sqrt{1 - 0.25}} = \frac{1}{\sqrt{0.75}} = \frac{1}{0.866} = 1.155 \tag{79}$$

for a length of:

$$L = \frac{L_p}{\gamma} = \frac{20.0 \text{ m}}{1.155} = 17.32 \text{ m} \tag{80}$$

It makes sense that humans made it through so much of their history without needing to understand special relativity: the impact is tiny until the speed being measured starts to reach a reasonable fraction of the speed of light. But what is a negligible impact in some contexts is absolutely critical in others. In the field of particle physics, for instance, scientists deal with particles that are *relativistic*: they move with some significant fraction of the speed of light on a daily basis. Scientific calculations would be very wrong if physicists did not take special relativity into account.

Special relativity forces us to make two difficult concessions. It requires us to accept that a distance (or length) that we measure depends on how fast we are moving with respect to the object. It also forces us to accept that the length of time that we measure depends on the speed that we are moving with respect to the action that we are measuring. In other words, our measurements of time and space are not absolute—they are relative. These two concepts, length contraction and time dilation, brought about an earth-shattering shift in how scientists thought about the natural world.

Absolute to Relative Choreographic Space

Pretend that aliens are flying on a spaceship past a performance that you have choreographed. They are moving at close to the speed of light. Using one of the movement phrases you developed in an earlier chapter, first design a version of your phrase that approximates what aliens would see from their spaceship. (Hint: you will need to compress all movements that occur along the direction that they travel.) Second, perform your movement phrase in a way that allows the aliens to come closer to seeing what would be seen by an audience on earth. (Hint: you will need to stretch your phrase out across the studio, along their path of travel.)

Drawing on physics for choreographic inspiration, as we ask you to do in this exercise, tests your understanding of relativity. It also injects a different kind of spatial thinking into your choreographic composition. Back on planet earth, choreographers tend to think of movement material as pliant putty to stretch, condense, fold, cut and paste, and braid by whatever means possible.

Choreographers experiment with space through formal manipulations of movement material. A choreographer could change the directions that a dancer faces within a phrase, for instance, or require the dancer to perform the same movements at a higher or lower level in relation to the floor. "Ironing out" a phrase means advancing all of the movements along a forward trajectory; "compressing" the phrase means staging the material within a much tighter area, which will change how the dancers perform the movement, and thus alter its look and feel. What happens if you perform the entire phrase while in contact with others? Or while moving along in a cluster? What if you inserted jumps and send that material flying across the stage? Choreographers look for spatial variations that give the composition texture and depth. The formations in which dancers perform the material can change the look and feel of the movement.

The ways that dance artists have thought about space have transformed significantly over the past hundred years. Generalizing about these changes is difficult, because many artists still use classic dance composition strategies that have been around for decades. But it is safe to say that, while early-twentieth-century Western ballet and modern dance emphasized creating *shapes* with dancers' bodies, contemporary dance frequently emphasizes *moving images*. These moving images remain in constant flux—as if mirroring patterns of thinking that never settle but grow increasingly fleeting and fragmentary. Think of changing cloud patterns as opposed to sculptures in stone. The transitions have become the form.

The very process by which choreographers research their formal choices has also changed. These aesthetic shifts are evident when we compare the language that artists from different time periods use to talk about their work. In her seminal book *The Art of Making Dances*, published in 1959, Doris Humphrey puts forward a theory of choreographic practice that is rooted in an overarching sense of order and symmetry. Humphrey's spatial design

focuses on creating bodily shapes according to classical principles of form. Areas on the proscenium stage dictate meaning: dancers performing material upstage will appear more godlike; dancers downstage will appear more human.[47] Humphrey's dance occurs in a Newtonian universe of absolute space and time.

In one way or another, contemporary dance has overthrown every aspect of these classical tenets. There are now far too many techniques and choreographic strategies to capture in a single book, as Humphrey tried to do. Working methods and physical practices, ways of performing, and the perspective of the audience are all open to question. According to the choreographer Jonathan Burrows, the only certainty is constant self-reflection. As he writes in *A Choreographer's Handbook*, published in 2010, imagining space requires questioning the means by which you arrive at that space: "Which way are you working? Which way do you want to work? Which way does your material allow you to work? ... What do you want from the body/What can it give you?"[48] Burrows's theory of choreography embraces a profusion of perspectives and choices that constantly undercut themselves. In opposition to Humphrey's ordered universe, space has become relative.

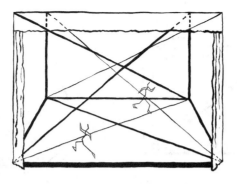

Stage Space

When a choreographer places choreographic material in a performance space he or she must think about other questions. From what angle does the choreographer wish the movement to be viewed? Does the composition dictate the spectator's focus at every moment, or does it give the viewer options as to where to look? How does the choreographer wish to define foreground and background? How will he or she fill the volume of an empty stage?

Stages come in a variety of shapes, each of which carries its own assumptions about how a piece will be performed. Concert dance is commonly presented on a proscenium stage, which separates the performers from the audience with an archway that frames the performance as if it were a picture. The audience faces the stage straight on. A strong sense of "foreground" and "background," as well as the obligation that the performers direct their performance toward the audience and not away from it, circumscribes the creator's choices.

Choreographers have challenged these limitations in many ways. Among other strategies, they have placed the performance in the round, created intimate performances for just one or two viewers, flipped the usual orientation of the theater by placing the audience on the stage, and even positioned the audience *under* the stage. Dance can also be presented outdoors or in other site-specific locations such as museums and art galleries.

Consider the radically different impact of even the most basic rearrangement on the audience's perspective, from the proscenium perspective in the drawing on the left to the dance in the round in the drawing on the right, below. History and culture weigh down these staging arrangements, which present creators with very different expectations and also offer radically different viewing experiences to audiences.

Good artists will rigorously question every aspect of their work, including the frame in which it is presented. The Congolese choreographer Faustin

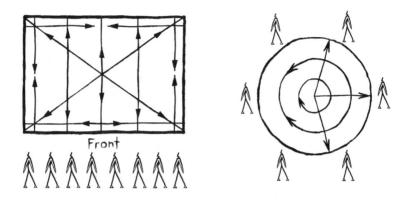

Linyekula reimagines the proscenium stage, which he describes as an expression of "the colonial relationship," by setting circle dances, a shape from his native Congolese dance.[49] Linyekula creates his work at his Studios Kabako, which he founded in the Democratic Republic of Congo in 2001. While he tours internationally to significant acclaim, he does not try to fit Central African culture into Western molds for contemporary performance. Nor does he represent Central Africa from the perspective of an outsider, as an expatriate might. Instead, his dances mirror the space and time of the Congo, a country whose people have experienced political upheaval and violence over the decades since his birth. Working inside the framing of the proscenium stage, Linyekula breaks that form with his choreographic patterns, thereby using his uncompromising point of view as a Congolese artist to defy the mores of the Western stage.

Perspective

Physics and dance share a preoccupation with perspective. Whether considering space on the concert dance stage, in a physics laboratory, or in outer space, the viewer's position will affect her or his interpretation of an event. The same movement phrase will look different depending on its arrangement on the stage and the audience's placement as spectators. For the person creating or performing the phrase, the perspective differs even more radically. Special relativity affirms that the viewer's point of view alters his or her experience of space: the measured length of an object will change with an observer's motion.

The innovative choreographer bends an audience's experience of space through movement, and the groundbreaking physicist asks others to process the anti-intuitive reality that length is not an unchanging constant. Both insist that the receiver question his or her position in relation to what is known. Both question what is known.

Could watching a dance that decentralizes the stage space offer a new way of understanding special relativity? Could an understanding of special relativity in turn open up new ways of viewing a dance? The conjunction of physics and dance holds the potential to raise new questions in both fields.

9. Time

Imagine the avant-garde theater director Robert Wilson preparing his next production. His 1976 breakout work, *Einstein on the Beach*, created in collaboration with the composer Philip Glass, dealt with special relativity and scientific progress. Unfolding over approximately five hours, *Einstein on the Beach* used dance, visual image, and music to investigate the effects of extended duration on the viewer's perception. Ever searching for the shock of the new—the greatest challenge to the avant-garde is to remain *avant*, continually innovating methods and meanings—Wilson has decided to take a radical turn in his latest work by casting fundamental particles in the principal roles. Today he is auditioning muons and top quarks.

Wilson leans back in his plush velvet seat in the darkened theater. His assistant leans over: "We're starting with muons." He nods. "Good. Bring them on." He stares at the blank stage. "What's happening?" he asks her, when nothing happens. She shrugs her shoulders, uncertain. "Next muon!" he barks. He stares at the empty stage. "Gone," she says simply. He breathes a quick snort of impatience. "Let's try this again." Through her headset, she invites the third muon onto the stage. "Muon, stay put!" he commands over the microphone. After a moment, he asks her, "Is it there?" They both stare at the stage questioningly.

An individual emerges from the back of the theater and stumbles through the darkness toward the tech table. It's the muon handler, a particle physicist. She murmurs in the assistant's ear. The assistant leans in toward Wilson: "Bob, it appears that muons have a lifespan of 10^{-6} seconds." He blinks. The muon handler speaks up: "If you wanted to spend more time with them, you would need to get them moving at close to the speed of light while you remained still. The effects of time dilation would make them live longer— for seconds even, if you got them going fast enough—from the perspective of your reference frame on earth."

For an artist who creates five-hour productions, this time frame is unfathomable. Wilson wonders what kind of movements can occur if the darn things keep disappearing, leaving aside that they are totally imperceptible to humans while they are alive. What's *presence* like at 10^{-6} seconds? Does a word like *alive* even apply? How does it feel to be a muon? ... and how would audience members know what they were looking at?

"Can these things be controlled?" he blurts out finally. The implications for his avant-garde theater are huge. He is sure he can make those muons hang around.

This is a fictional thought experiment, of course—no director could audition fundamental particles, or stage a production at light speed. We are using the scenario to bring our two fields together, to reveal basic assumptions about time, perspective, and existence. Physicists who use muons in their research actually share Wilson's problem: how to make muons live longer, in order to do experiments with them. Live performance is a petri dish in which to experiment with time, something physicists do in their laboratories under very different circumstances.

Our separation of the discussion of space and time into two chapters has been artificial, in fact. Wilson's auditioning muons would not only appear to live longer the faster they traveled; the distances they traveled would also change, based on the effects of special relativity. In modern physics, it is difficult to talk about time or space in isolation. A more appropriate term to describe current understanding of reality is *spacetime*. Physicists in the early twentieth century joined these two words to signify the profound interrelationship of space, time, and objects in motion.

Choreographers of the twentieth century, like physicists, also began to probe the interconnectedness of space and time. Through choreographic practice, dance artists have learned that manipulating time involves manipulating space. Various combinations of speed, duration, rhythm, and stillness will alter both a viewer's spatial and temporal experience.

In this chapter, we look at some of the ways that physicists and dance artists work with time, all the while keeping in mind the conceptual interdependence of time and space that grounds both disciplines.

Special Relativity: Time Dilation

When we presented the two postulates of special relativity in the previous chapter, we assumed that the physics of all inertial, or nonaccelerating, reference frames is identical and that the speed of light is always the same in a vacuum. This led to the startling conclusion that the measurement of an object's length depends on how fast the observer is moving relative to the object. Objects can only be seen to contract in space if the observer is moving at relative speeds close to the speed of light. What about the amount of time that an event occupies? Special relativity unmoors time as fundamentally as it unmoors space.

Let's think about measuring the time it takes for an event, such as a dance performance, to unfold. Our event needs a well-defined starting point and ending point to enable us to measure its duration. Imagine that two observers in two different inertial reference frames are each measuring the performance time. We can write down a formula that governs the relative times the two observers will measure for the event.

Just as the proper length L_p belonged to the observer at rest with respect to the object, the proper time t_p is measured by the observer at rest with respect to the event. The other time t is measured by the observer whose reference frame is in motion with respect to the event. Here is the formula for time dilation:

$$\Delta t = \Delta t_p\, \gamma \qquad (81)$$

with

$$\gamma = \frac{1}{\sqrt{1 - \frac{v^2}{c^2}}} \qquad (82)$$

We use the symbol Δ to signify "difference." We are measuring the difference

$$\Delta t = final\ time\ -\ initial\ time = total\ time\ of\ an\ event \qquad (83)$$

Since we cannot travel faster than the speed of light, $v < c$ and γ will be a number greater than 1. Looking at formula 81 above and applying this constraint, we can see that the event time measured by the observer in motion will always be greater than the time measured by the observer in the same frame as the event.

To clarify this and as practice in assigning Δt and Δt_p to a physical scenario, let's return to the example in the previous chapter: a flashlight performance occurring in a dance studio on a moving train.

The observers in the studio measure the time that it takes for the light to travel from the flashlight up to the ceiling and then back again. Because they are not moving with respect to their experimental apparatus (flashlight and mirror), they measure Δt_p, or "proper time."

The audience on the side of the road watches this performance on the train as it passes in front of them. They also measure the time for the light to travel from the flashlight, up to the mirror, and then back down again. Their time is associated with the no-subscript Δt in the formula above. As we saw with our length contraction exercise, the times measured by the two observers, Δt_p and Δt, will be quite close to each other when the velocity of the train is far below the speed of light. We will begin to see Δt become significantly larger than Δt_p as the relative speed difference between the two reference frames increases.

"Wait!" you might be thinking. "If the performers are on the moving train, how can they be the ones measuring proper time? The observers on the side of the tracks are the ones standing still." But remember that in order to decide which observer measures a proper length or proper time, we must identify the observer who is at rest with respect to the object or event being measured. Everything is, according to something or someone else, hurtling through space. There is no such thing as "completely still" in an absolute sense.

We saw in the previous chapter that if two reference frames have a relative motion of $0.5c$ (or 50% of the speed of light), then γ will have a value of approximately 1.155. We repeat the calculation here:

$$\gamma = \frac{1}{\sqrt{1 - \frac{v^2}{c^2}}} \quad = \quad \frac{1}{\sqrt{1 - \frac{(0.50c)^2}{c^2}}} = \frac{1}{\sqrt{1 - 0.25}}$$

$$= \frac{1}{\sqrt{0.75}} = \frac{1}{0.866} = 1.155 \tag{84}$$

We can make our example concrete by choosing an amount of time that the performers measure for the event. If they measure the performance to take one second, we would set Δt_p to equal one second. The performers are at rest with respect to the performance, so they are measuring the proper time. To observers moving at $0.5c$ with respect to the performance, the time Δt measured would be

$$\Delta t = \Delta t_p \, \gamma = (1 \text{ s})(1.155) = 1.155 \text{ s} \tag{85}$$

The closer the train gets to the speed of light with respect to the ground, the bigger the difference between the times reported by the performers and the observers would become. Which time is correct? Both! Time is no more absolute than distance is.

Length contraction and time dilation each depend on γ, which encodes the relative velocity of two reference frames. For a pair of observers, the effects of relativity on time and space are inexorably intertwined through this variable. Once we understand that time and space are linked in this way, it does not make sense to think about them independently. At the turn of the twentieth century, physicists began putting space and time together in their calculations and their speech. Spacetime was born.

Spacetime

Spacetime is not solely a mathematical conceit proposed by modern physicists; space and time are deeply connected in our everyday experience, too. In using movement to manipulate time and space, no art form shows this interrelationship better than dance.

Consider the ideas of a pioneering physicist and a paradigm-busting choreographer side by side. The following proposition is often used to summarize Einstein's thesis in his 1917 theory of general relativity:

Matter tells spacetime how to curve, and curved spacetime tells matter how to move.[50]

And here is the choreographer Merce Cunningham, writing in 1952:

The fortunate thing in dancing is that space and time cannot be disconnected, and everyone can see and understand that.[51]

141

Cunningham's words tell us that the movements of a dancer make the interconnectedness of space and time visible. Einstein's theory expresses a similar idea to Cunningham's, only on a cosmic scale.

For both thinkers, space and time are conceptually wedded. But to perceive the curvature of spacetime, Einstein needed to think in gigantic terms, using incredibly large bodies in space. Notice here the importance of scale on perceptibility and measurability: often we need to move into extreme conditions in order to learn something more about a phenomenon. This is a theme that will recur throughout this chapter.

It is worth examining the theories of Cunningham and Einstein more closely—one expressed choreographically, the other scientifically—to better understand their implications for time.

Cunningham's "Space, Time and Dance"

Much of Cunningham's theory is expressed in his dances, created over seven decades of making work. He also wrote about dance, albeit infrequently, and in his writing we can find clues that help us to better understand his choreography. Early on in his short essay "Space, Time and Dance," first published in 1952, Cunningham tips dance aesthetics away from Laban's Euclidean universe into a more relativistic mode. In a neat inversion, he begins the essay by deconstructing the conventional use of space in concert dance, and he ends by proposing some choreographic innovations in terms of time.

In order to clear a pathway for his new aesthetic, Cunningham had to first challenge historical precedents. (This should sound familiar: remember that Einstein, too, discredited prevailing assumptions in physics, in order to propose his new theory of special relativity.) Cunningham starts the essay by challenging the conventional uses of space in concert dance. In particular, he describes the linearity of classical ballet and the "lumpy" formations that he observes in much of German expressionist dance and American modern dance. He dismisses these usual treatments of space as possessing an inherently static, inactive relationship to time.

Cunningham does considerably more aesthetic path clearing in the essay. He tackles time by calling into question the imperative to include dramatic climaxes in a performance. Taking issue with the dynamic phrasing of much concert dance, he narrows his focus to the basic structure of a movement phrase. To Cunningham, the expectation that a phrase will contain a rise, climax, and fall falsely suggests a "crisis to which one goes and then in some way retreats from." But *climax* is made meaningless, in his mind, by the fact that there are so many small crises—and thus climaxes—in life. Life goes on, in a continuous series of discrete actions. "Climax is for those who are swept by New Year's Eve," he writes.[52] Cunningham is calling for dances that feel more like life.

Having argued against these various aesthetic conventions, Cunningham can propose his new theory. His radical suggestion is to experiment with structures of time so as to reinvent the organization of movement in space. He writes: "More freeing into space ... would be a formal structure based on time. ... If one can think of the structure as a space of time in which anything can happen in any sequence of movement event, and any length of stillness can take place, then ... counting is an aid toward freedom, rather than a discipline toward mechanization."[53]

In thinking of choreographic structure as a "space of time," Cunningham frees movement from its temporal dependence on music. His movement could follow its own internal meters and stillnesses and fall into any sequential order. When he describes counting as "an aid toward freedom," he means that rigorous time structures can give movement an internal cohesiveness, independent of the structures dictated by the music. This new way of thinking about choreographic composition informs a number of Cunningham's aesthetic innovations, which he pioneered in collaboration with his lifelong partner, John Cage. Three innovations in particular contribute to the significance of Cunningham's work.

The first innovation is his belief in the autonomy of dance and music. A choreographer and a composer could create independently and still present their work simultaneously. Freed from the obligation to illustrate music, the dance could carry its own innate expressiveness. As Cunningham explains, this arrangement allowed the connection between the dance and the music to be "one of individual autonomy connected at structural points."[54] Any links between the dance and the music would arise serendipitously, pieced together in the minds of the observers in the moment of performance.

This new freedom required that Cunningham invent time structures to support the dance. One of his leading dancers from the 1950s, Carolyn Brown, describes compositional processes in which Cunningham precisely varied the duration of movement sequences: a phrase might last a minute, or thirty seconds, or fifteen seconds. This simple manipulation of time changed the movement quality, as well as the dancers' pathways through space.[55] Cunningham's dancers internalized his rhythms, learning to memorize entire dances as an intricate series of beats. As a result, they appear to be concentrating intently, as if solving complex math problems while exhibiting perfect modern-dance technique.

In a second innovation, Cunningham used chance operations to create his dances. With a toss of a coin, papers pulled from a hat, or other devices, he gave over to chance certain compositional choices, such as the duration of a movement or phrase and its spatial patterning. The process bypassed the creator as artist and turned over aspects of the decision making to nature. After Cunningham used chance to assemble the steps, the choreography remained set; it was never improvised. Cunningham's use of chance procedures, or "indeterminacy," created surprising transitions, timings, and spatial organizations.

Lastly, in a third innovation stemming from his use of chance operations, Cunningham altered the way dancers used the stage space.[56] No longer were there such things as "important" events that had to happen center stage for all viewers to see. Events could erupt onstage anywhere and at any time. His 1952 *Suite by Chance* (a piece he later expanded and renamed *Suite for Five*) is the first dance in which he applied chance procedures throughout the development process. His subtle experimentation with space becomes clear as his movement phrases set dancers roving unpredictably around the stage. His performers face every direction, rather than either toward the audience or upstage, away from the viewers. Formations surface and disappear throughout the stage space. Dance artists call this a "decentralized" use of the stage: no one point or facing is more important than any other.

The nascent choreographic theory that Cunningham expressed in his 1952 essay and in the creation of *Suite by Chance* held true for the duration of his career. He understood that by manipulating time, he could alter space. This interdependence of time and space arguably lies at the core of his groundbreaking choreographic practice.

These innovations might seem like choreographers' shop talk until you watch a Cunningham dance. Activities occur, stillness offsets movement, and dancers pop up asynchronously in solos, duets, or trios. The images created by the dance exist fleetingly, one dissolving into the next. It can be difficult to tell when a Cunningham dance is nearing the end. A sense of continuous motion pervades his dances, as if the dance were not limited to the curtain-up/curtain-down timing of theatrical performance, but occurred in perpetuity, as in nature. And just as in nature—think of planetary masses or stars—the dancers appear to carve out space and time as they go.

General Relativity

Cunningham's experiments with space and time changed how people saw dance. What are the implications of Einstein's work? His theory of special relativity implied that time and distance were relative and linked. He followed these ideas with a theory of general relativity, which resulted from his effort to make a consistent model of nature that included both special relativity and gravity. General relativity gives us another example of the necessity of spacetime.

Recall the discussion of the physics of gravity in Chapter 1. There gravity was treated as a force. It is always attractive. As posited in Newton's Law of Universal Gravitation, the strength of the force of gravity depends on the masses of the objects interacting as well as the distance between them:

$$F_G = \frac{GMm}{r^2} \tag{86}$$

where G is Newton's gravitational constant, M and m are the two masses experiencing the gravitational attraction, and r is the distance between their centers of mass.

In general relativity, Einstein abandoned the idea of gravity as a force altogether, and instead proposed a fabric of spacetime. Mass curves spacetime. The larger the mass, the greater the curvature. The motion of objects

144

can be explained by how the warped spacetime incites them to move. In this new picture, gravity is a manifestation of the curvature of spacetime.

This is a beautiful idea that can be illustrated as masses forming contour maps on a surface, with objects rolling down indentations providing the impetus for gravity.

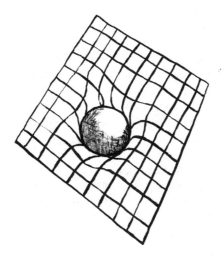

Several revolutionary implications follow from Einstein's general relativity. One implication is the *principle of equivalence*: from the physics perspective, an accelerating reference frame is identical to a reference frame in a gravitational field. To put it another way, these two experiences would feel identical:

1) You are standing in a dance studio on the surface of the earth and thus experience an acceleration due to gravity at 9.8 m/s^2, which is perfectly counteracted by the floor that holds you up. Bending your knees, you jump.

2) You are standing in a dance studio on a rocketship way out in space, accelerating at a rate of 9.8 m/s^2 in the direction you would define as "up," toward the ceiling of the studio/ship. Bending your knees, you jump.

You might imagine that the experience of jumping in these two scenarios would feel different, or be different in some measurable way. The first sensation and resulting motion is due to your proximity to a large mass, which is pulling you in with a gravitational force. The second is due to the push of an engine propelling you in some direction. But general relativity tells us that you would not be able to differentiate between being on a planet's surface and being carried in a rocket ship accelerating at a rate that matches the planet's gravitational acceleration. This is Einstein's principle of equivalence within general relativity.

General relativity also tells us that the strength of a gravitational field affects the passage of time. Remember that special relativity taught us that if we have two observers with completely accurate clocks, these clocks can run at different paces if the two observers are in motion with respect to each other. The same can be said of two observers in different gravitational fields. For example, an observer on the surface of planet earth is in a stronger gravitational field than an observer on the moon. Because of this, the observer's clock on the earth's surface will run at a slower rate than the clock on the moon.

Should we all, then, move into the earth's deepest valley to lengthen our lives? Remember that a significant difference in time dilation occurs only when extreme differences exist in the relative velocity between two observers. Similarly, a significant impact in the rate of the passage of time due to gravitational field differences occurs only when the strengths of the gravitational fields differ dramatically. And just as a human cannot easily accelerate to reach anywhere near the speed of light, so that observer cannot get close to a huge gravitational field—such as near a black hole—in order to experience these differences. The tests of general relativity, like the tests of special relativity, are therefore constrained to phenomena that fall outside our unaided perception—phenomena such as the bending of light rays, the behavior of fundamental particles, and gravitational waves. Don't think, however, that special and general relativity bear no relation to your everyday lives: global positioning systems (GPS), for example, must correct for the effects of both.

Extended Duration

To develop his theories of time, Einstein had to imagine extreme conditions: light speeds, vacuums, the yawning void of outer space. Artists have also tested time by going to extremes, experimenting with everything from extending the duration of a performance to slowing down a single blink of a performer's eye.

Whereas Merce Cunningham did away with dramatic climaxes to more closely represent the activity of everyday life, the reverse occurs in extreme-duration works. Artists can eke drama out of the most ordinary actions by stretching a composition in time. One example of such work is *The Artist Is Present*, created and performed in 2010 by Marina Abramović, in which she sat still and silent at a small table in the atrium of the Museum of Modern Art while visitors were invited to sit across from her and gaze into her eyes. The table and chairs, coupled with a rectangle marked out with tape on the floor, might seem like an unremarkable set up, were it not for the duration of the performance. Abramović sat for a total of 736.5 hours over a period of three months during the museum's open hours. This feat of endurance produced in viewers a combination of empathy and morbid fascination—not many people would put themselves into such a scenario for the sake of art or anything else.

In playing with extreme durations, this type of performance art undeniably impacts human perception. Observation over time alters what we perceive.

To ease into the topic of duration, try a movement study that investigates perception. Set up a "performance" in which one person watches another person who is standing completely still. The person playing the role of "performer" should avoid making eye contact.

Set the clock for three minutes.

. . .

. . .

. . .

Each of you take note of what you experienced during those three minutes. Both watcher and performer may have found themselves noticing minute details—the sound in the room, the twitch of a cheek, a subtle shift from one foot to the other. As the time progressed, the observer may have found her- or himself noting personal characteristics of the performer. At one moment the person cannot conceal impatience; at another thoughts flit transparently across the performer's face, discomfort and acceptance following in quick succession.

Stillness in performance allows for qualities to rise to the surface which would not be apparent if the performers were perpetually moving, or moving at an everyday rate. Remaining still over a period of time distills a performance down to presence—a deceptively simple state of being in time and space that allows dramatic meanings to arise in the exchange between observer and doer.

This three-minute mini-performance was drawn from an audition for one of Robert Wilson's productions. While he works with highly skilled performers, how compellingly they can simply *be* while being watched is as important to Wilson as how they move.

Wilson uses stillness and slowness to develop his theatrical scenes at an extremely slow pace. In his five-hour opera *Einstein on the Beach*, scenes unfold at a crawl. His performers execute deliberately stylized movements, in which every gesture is controlled. Their sculptural quality amplifies the actions that occur—a stiff walk, a nervous arm tic, a flick of a head. An Einstein-like character appears, playing the violin. Einstein here is a symbol of scientific advancement, fittingly subsumed and abstracted in the production's formal experimentation with space and time.

Manipulations of time drive the staging. In one vivid scene, "Train," the dancer Lucinda Childs, who was also a collaborator and choreographer for the production, skips forward and back along a diagonal in a repeating phrase of quick steps and hand signals set to a rhythm that mirrors Glass's music. Her fellow performers execute highly abstracted, task-like gestures

147

at varying tempos. One appears to be operating a machine of buttons and pulleys, another reads the newspaper, another solves a math problem. Their pathways through the space differ: one remains in place, another moves only in a series of right angles. In the background, behind this multilayered choreographic composition, a large train looms into the frame from the right and advances across the stage. The train is another dancer, adding another rhythm.

The opera contains no single dramatic climax. In lieu of the conventional theatrical arc, *Einstein on the Beach* accumulates images and movements, spaces and times that reveal the impact of scientific and technological progress on human experience. However subtly or violently, both scientific and aesthetic invention transform our experience of time.

The Lifetime of Muons

Moving from avant-garde theater to one more scientific example of time's ability to bend according to the circumstances, let's return to particle physics and think about time dilation in the lifetime of particles.

In the early twentieth century, scientists learned that particles from the universe were bombarding the earth's atmosphere. The collisions created cascades of other particles showering down from the sky that were named *cosmic rays*. Many new particles were discovered in cosmic ray showers. One particle, discovered in 1936 and eventually given the name "muon" (pronounced mew-on) seemed particularly anomalous. Muons were an unstable, heavier cousin of the electron that could not be accounted for by the theories of the day. They had the dubious distinction of prompting the physicist I. I. Rabi to complain, "Who ordered that?"

Muons have an average lifetime of .0000022 seconds, or 2.2 microseconds. Any given muon could exist for a longer or shorter period, but in a large group of muons, the average time it takes for one to cease to exist and decay into other particles will be 2.2 microseconds. We know roughly where in the upper atmosphere these muons are created in cosmic ray showers. Here's the rub: we see many of them at the surface of the earth, but by earthbound calculations most of them should not live long enough to reach us. The number of muons that reach the ground defies the calculations of Newtonian physics. Special relativity, again, comes to the rescue.

Think about the time a relativistic (*very* quickly moving) muon speeding through our atmosphere would measure for its existence. We can compare the muon's measurement of its own lifetime with what an observer on earth would measure of that lifetime. If we set up our scenario with the equation

$$\Delta t = \Delta t_p \, \gamma \tag{87}$$

who can claim to measure the proper time, Δt_p? The event in question is the length of time that the muon exists. It is at rest with respect to itself, so in measuring the span of its life, the muon can claim the proper time.

We shall assume that this muon exists for exactly the average amount of time that a muon tends to exist, 2.2 microseconds. If the muon is traveling at 90% of the speed of light, what muon lifetime would the person on earth measure? We can first calculate γ using the difference of $0.9c$ between the reference frames:

$$v = 0.9c \tag{88}$$

$$\gamma = \frac{1}{\sqrt{1 - \frac{v^2}{c^2}}} = \frac{1}{\sqrt{1 - \frac{(0.90c)^2}{c^2}}} = \frac{1}{\sqrt{1 - 0.81}} = \frac{1}{\sqrt{0.19}} = \frac{1}{0.4359} = 2.294 \tag{89}$$

Putting these values into our equation to calculate the time measured by an observer on earth, we can see that there is a huge discrepancy in terms of the two measurements of lifetime:

$$\Delta t = (2.2 \ \mu s)(2.294) = 5.047 \ \mu s \tag{90}$$

The muon measured its lifetime at 2.2 microseconds. But to observers on earth, the muon seems to have stretched its lifetime out to more than double the average, 5.047 microseconds. And if we were to make the measurements for many muons, we would quickly see that the average that we measure is much higher than 2.2 microseconds.

To reconcile this contradiction, we can think about this muon scenario from the perspective of length contraction. First we need to form clear questions: How far has the muon traveled in its life from its own perspective? And what is this length from the perspective of somebody who is standing on the earth? Let's assume that the muon travels straight down from our upper atmosphere toward the center of the earth. The formula that lets us work with lengths when our reference frames are moving at relativistic speeds with respect to each other is

$$L = \frac{L_p}{\gamma} \tag{91}$$

Does the muon or the observer on earth measure L_p, the proper length, of the distance traveled through the atmosphere? Since observers on earth are the ones at rest with respect to the atmosphere, they are the ones who measure the proper length. The muon measures L.

Let's assume that the muon is moving with respect to the earth's surface at 90% of the speed of light, or $0.9c$. The relative velocity between the reference frame of observers on the earth and the frame of the muon is therefore the same value that we used in the previous example, which will give us a γ value of 2.294.

If the muon is traveling at $0.9c$, and it has 2.2 microseconds to travel, it measures the distance of travel by multiplying its velocity by its time:

$$L = 0.9c \times 2.2 \times 10^{-6} \text{ s} = 0.9 \times (2.99 \times 10^8 \text{ m/s}) \times (2.2 \times 10^{-6} \text{ s}) = 592 \text{ m}$$
$$(92)$$

We can enter the 592 m that the muon measures as our "L" in the formula above. If we solve for L_p, we will know the distance the muon has moved through the atmosphere before it decays, from the observers on the ground:

$$L = \frac{L_p}{\gamma} \rightarrow L_p = L\gamma \qquad (93)$$

We have calculated that L = 592 m and γ = 2.294. Our equation therefore looks like this:

$$L_p = (592 \text{ m})(2.294) = 1358 \text{ m} \qquad (94)$$

To summarize what we learn from these two calculations: the muon would claim, "I lived 2.2 microseconds and traveled 592 meters." And the observer on earth would claim, "The muon lived 5.047 microseconds and traveled 1358 meters." Each would be telling the truth, because both are making accurate measurements. Time and distance are relative, depending on how quickly something is moving with respect to what is being measured.

Some of the muons created by cosmic rays in our upper atmosphere are traveling even faster than the $0.9c$ that we assumed in the calculation. They can reach speeds of $0.999c$ and above. Their lifetimes are therefore stretched even longer from the observer's perspective (and the distance is squeezed even smaller from their perspective). As a result, many more muons reach the earth's surface than we would predict without taking into account the length contraction and time dilation of special relativity. In particle accelerator experiments we can apply these principles of special relativity to keep short-lived particles around longer, simply by speeding them up. Time dilation and length contraction conspire to give us more time to do experiments with unstable particles.

If we can extend the life of a muon in this way, could we also extend the life of a person? Absolutely. The faster we get someone going with respect to us, the slower the person's time will run with respect to our clock. However, the person would still experience time passing at the same rate in his or her own life, as opposed to experiencing extra time. The person would need to be moving near the speed of light with respect to you before your two rates of aging became significantly different. Since scientists do not yet have a simple, safe way of accelerating people to near the speed of light, they concentrate on experimenting with the lifetimes of particles in their laboratories.

Rates of Change

Try to imagine those muons auditioning for Robert Wilson, with his love of stillness over time . . . either none would get the job, or Wilson would need to adjust his parameters for "presence" in performance. The lifetime of a muon does not correlate to the time required for our perceptual faculties to function: we can perceive very little that occurs in the space of 2.2 microseconds. To complicate this comparison further, particles do not age. A muon pops into existence and then it disappears and other particles exist in its place. There is no process of living, aging, and dying. There is no internal rate of change.

For physicists, time is a measured construct in which change may (or may not) occur—but change is not a requirement of the definition. In contrast, in both our everyday experience and in art making, time by definition implies change. Humans are born, age, and die; live performance mirrors this progression. Performances have beginnings and endings, and in between the expectation is that some kind of change will take place.

If change must occur, how do artists create it? Sometimes change in performance happens through the classic dramatic structure of build up, climax, and resolution. For artists who resist this structure, change can mean prolonged exposure to an activity: a window opens, things happen, the window closes. The impression (a sleight-of-hand) is that the activities continue with or without the viewer present. Change can also happen through prolonged exposure to *no* change. Spending an hour watching performers engaged in a single action transforms the viewer's perception—not unlike the childhood game of repeating a word over and over. Just as the word can lose its original meaning and offer up new sounds, the same movement viewed over time can become strange. Viewers will begin to notice new dimensions to a previously familiar action.

The driving questions for any creator of live performance are: How long should things last? what should change? and how long should it take for things to transform? These questions lead to rates of change, which deal with how quickly or slowly, abruptly or imperceptibly a composition changes over time. Manipulating rates of change can generate narrative—a term we use in performance to refer to the stories, or series of actions and relationships, that develop over time, and that may be literal or abstract. To better understand the interplay between rates of change and the unfolding story, you can try a movement exercise.

This exercise requires two performers and one observer. The first performer will stand still, somewhere within the performance space, keeping his or her eyes focused toward the horizon line. The second performer should be positioned approximately 2.5 m away, facing the first performer. The second performer will walk slowly toward the first performer, moving at a constant rate and using the entire three minutes to reach the other. The observer should simply watch. Set the clock for three minutes, and go.

. . .

. . .

. . .

This study is abstract, in the sense that the performers are not playing specific characters, nor are their life trajectories spelled out. We can still see a narrative, however, in these simple elements. Two people sharing a space, the carefully modulated rate of change, and the pathway of one performer heading toward the other imply a dramatic relationship.

You may have noticed yourself asking questions during the three minutes: Why is one advancing on another? Why does the one standing not respond to another? What does each person want? Did the walker make contact with the stander at the three-minute mark, or redirect and pass the other by? In our instructions we left the scenario's conclusion intentionally vague. Even knowing the instructions, the set-up creates a sense of mystery.

The relationship between the two performers remains unclear, and it shifts as the directive progresses. Is this a situation of threat or aid? You might read a different scenario with each second.

The renowned dance artists Eiko & Koma create such ambiguous narratives, stretched in time. Disciplining their bodies to fully actualize each and every second of motion, Eiko & Koma shift their images at a pace "suggestive of geologic scales."[57] Growing up in Japan in the 1960s and 1970s, they studied butoh with a master teacher, Kazuo Ohno. His influence is visible in their use of time, from their pervasive aesthetic of slowness to their attention to the human condition. In motion, their bodies appear to be simultaneously hollow shells and brimming with life. Performing outdoors in rivers and indoors in sculptural installations of feathers, dirt, and leaves, among many other environments, Eiko & Koma erase the boundaries between humans and nature, the artificial and the humanmade.[58]

In their 1989 piece *Rust*, Eiko & Koma are seen naked and upside-down, with their heads on the floor and their legs pressed up against a chain-link fence. During the performance, they writhe slowly along the links, disconnected from each other. Koma's approach seems to threaten Eiko, but when he reaches her, the story changes and he moves underneath her, supporting her body as if he were a bench or a bed. The visual design they create with their bodies generates an ambiguous drama. How they got there is unclear.

Eiko & Koma performances defy performance genres: Are they installations, sculptures, or dance or theater works? Each of these genres carries expectations about the audience's experience of time. While many installations or sculptures allow for extended viewing and permit the viewer to walk around the art, dance and theater frequently fix the audience's viewing position and average one to two hours in length. Eiko & Koma's attention to time, materialized in movement, dissolves these disciplinary boundaries. *Rust*, for example, is presented on a stage, but it offers up sculpture-time. This mixing of format and temporal modes gives their work a kaleidoscopic quality. Viewed at one moment, their bodies appear to be all lines and figures. Viewed a minute later, these same bodies appear to be celestial, recovering from a fall.

The depth of their inquiry into time and space has led Eiko to note, "Space on stage is brushed by time," a thought that sounds like an articulation of spacetime expressed through a life lived in performance.[59]

Gravitational Waves

One of the predictions of general relativity is that masses will emit gravitational waves—a stretching and compressing of spacetime itself. If a gravitational wave were to pass through you, your body would distort as if it were caught in an oscillating circus mirror that changes your shape. For almost a hundred years, gravitational waves were a prediction that could not be validated by experiment. The amount of the stretching and compressing was far too small for scientists' most sensitive devices to detect until they developed the Laser Interferometer Gravitational-Wave Observatory (LIGO) Experiment. Upgrades and fine-tuning over decades of research enabled LIGO to achieve the necessary sensitivity.

For gravitational waves to be detected, two criteria are necessary: First, there must be a huge amount of energy in the gravitational wave so that it will register on the instruments. A candidate for such a wave would be one produced by a merger of two orbiting black holes somewhere in the universe. There would be a tremendous amount of energy released as the two fused into a single black hole.

The second criterion is an instrument so sensitive that it can sense the stretching or compressing of spacetime by an amount smaller than the width of a proton. LIGO accomplishes this by sending light on paths of 4 km in two distinct directions and bouncing it back and forth between mirrors. Observers in the lab are able to detect if the paths of the light get out of sync due to minuscule stretches or compressions of the distances between mirrors.

In September 2015 the LIGO experimenters discovered what they described as ripples in the fabric of spacetime from what they deduced was a merger of two massive black holes that had occurred a little over a billion years ago. Since that time, they have had more gravitational wave sightings. It took over one hundred years for this prediction from general relativity to be discovered. It was a phenomenon that even Einstein—who made the prediction—thought would be impossible to detect. As scientists are now able to "listen" to gravitational waves they have a new tool to employ when probing the universe. LIGO collaborators and cosmology theorists received the 2017 Nobel Prize in Physics for this discovery.

Ghostcatching

Physics and dance share the singular problem of our universe: time moves in one direction. Events that occur can never be repeated exactly. A detector captures the collision of two black holes as an abnormal frequency—a cosmic blip, like the notation for a billion-year-old dance. A similar challenge comes from trying to re-create the dancing of Master Juba, a virtuosic African American dancer of the 1840s, through the writing and testimonies of those who saw him perform. Dance historians reconstruct events through their traces in the archive, just as physicists do when they interpret data from detectors. The reverberations of a past event may be felt today, but the event will never (indeed can never) occur again in the same exact form.

No moving image of Master Juba exists. But starting with the rise of film and then video in the twentieth century, the great dancers have been captured on screen—at least, those fortunate enough to be filmed. A collection of moving image technologies has emerged over the past one hundred and twenty-five years that has helped scholars to augment the dance archive: from film and video to smartphones and motion-capture laboratories, these introduce the ability to freeze, rewind, jump-cut, and slow down time to excruciating extremes. They can even record the essence of human movement, devoid of the human body.

Motion capture, a more recent computer technology, generates another kind of dance replication: an image that is not two-dimensional but three. Such a system produced *Ghostcatching*, a seminal work of digital art created collaboratively in 1999 by Paul Kaiser, Shelley Eshkar, and the dancer-choreographer Bill T. Jones. Motion-capture systems track sensors placed on the human body. With the data from each sensor, the computer assembles a visual representation of the movement that can then be applied in a number of disciplines, from medicine to digital animation and other works of art. Over a number of research sessions, Jones was suited up with sensors and then improvised various dance movements. Together, Jones, Kaiser and Eshkar refined the phrases, which Kaiser and Eshkar edited and treated with hand-drawn lines. The figural images that appear in the final video proliferate and recombine. The background is black and abstract, not the real world, with lyrical lines that both form Jones's figures and create layers, like a sediment through which he moves.

Ghostcatching unquestionably experiments with space, but the formal play with time most speaks to our purposes here. The video begins with what Kaiser has called an "ancestral figure," the creator who spawns all the other dancers that appear, all of whom are hand-drawn versions of Jones.[60] The movement phrases have been cut up and remixed from Jones's live improvisation. Through digital art, the creators craft altogether new characters and a new narrative.

Many dance writers, including Jones, have noted the absence of sweat in *Ghostcatching*. Motion capture, as Jones has observed, takes away dance's hard-won ephemerality.[61] The system turns movement into a series of data points that describe location and speed. But there is something missing in what the technology records, too. The figures hop, stretch, and crawl under alien physical conditions—their actions are too light, too buoyant. Theirs are not bodies on earth, but bodies in digital ether. Kaiser and Eshkar were concerned with intermedia translation and drawing, more so than with preserving Jones's exact performance. Their digitally rendered line drawings create a compelling new art while losing what the dance scholar Ann Dils has described as Jones's "animus"—his life force.[62]

Only in the sound do you feel his animus. Jones hums, narrates fragments of stories, and sings children's songs that evoke spirituals. The recordings capture the physics of sound on earth: Jones's vocal folds vibrate, and that vibration pushes the air out into high and low density patterns that carry through the air on sound waves. His vocalizations contain the human quality missing in the moving image. Jones's voice is the true ghost of *Ghostcatching*, as opposed to the traces of his motion that end up on the screen, because something more familiar to lived experience registers—something more true to the human body in its engagement with physical forces. Complicating the digitization, the earthly physics expresses the African American history and identity that Jones never leaves far behind. Take away that physics, and not only Jones's animus but also human history itself appears to evaporate.

Consider the gap between an event and its record as a kind of ghostcatching: what is imprinted in the archive, and what is left behind? What "animus" do scientists miss, on a much vaster scale, when they read data on the motions of planets and stars that occurred billions of years ago? How can we think about *liveness* and *life force* as bridges between humanistic and scientific inquiry?

What other kinds of questions should we be asking?

The Story of the Universe

Artists can tell stories, thereby shaping time. But how do we craft the story of the universe? Was there a beginning and will there be an end, or are we in the midst of a cycle that has always been repeating and will repeat for all time?

In physics, the current view of the universe is that it began with a bang almost 14 billion years ago. Scientists do not know the origin of the bang, and they do not know if it was the first bang that ever happened. More broadly, they also do not know whether humans inhabit the only universe that has ever existed, or even if other universes currently exist. Strong evidence leads scientists to believe that the earth is approximately 4.5 billion years old and that it will be engulfed by the sun within a few billion more years. (That leaves plenty of time for humanity to figure out intergalactic travel if we manage to continue scientific exploration.) As we reported in Chapter 7, scientists have evidence that our universe is expanding at an accelerating rate, which means that if there is an end to the universe, it will probably be cold and dark and difficult to define.

We can peer back in time by looking at objects farther and farther away from us. When we look up in the sky and see the moon, we are not seeing the moon as it exists now—we are seeing an image a little over a second old. It takes that much time for reflected light leaving the surface of the moon to reach our eyes on earth. The image of the sun we see in the sky is eight minutes old. The sun is farther from us than the moon, so it takes an additional amount of time for its light to reach our eyes. The North Star, Polaris, is several hundred light years away from us; the image of the star that we see at night is several hundred years old. If we look farther out into space we are peering even farther back in time. Assuming that we can account for disruptions in the light as it travels to us, we can watch the history of our universe unfold. And modern telescopes are certainly much more sensitive than human eyes. Some of our instruments allow us to access light that set off on its journey toward us billions of years ago.

Our understanding of the life cycle of the universe continues to evolve as our tools mature, and the story is certainly not over.

Perception

Both artists and physicists create structures that help them listen intently to the rhythms of the natural world. Whether through a physics experiment that relies on the lifetime of muons or a performance that holds the spectators' attention to the minutiae of minutes passing, we struggle to perceive the time and space in which we live.

Trying to fathom time in physics and dance raises more questions than it answers. How do the different ways of knowing time and space inform our notions of truth? How do cultural forces affect our perception? Do other temporal or spatial dimensions exist? How can we perceive them? How did this massive performance that is our experience in this universe begin, and in what way will our cosmos change? And (thinking like a choreographer or physicist) at what rate? What further assumptions are we carrying with us and how can we challenge them?

We leave you with these questions to contemplate. The great science-art collaborations of the twenty-first century will conjoin forms of reasoning— aesthetic, mathematical, scientific, and embodied—in order to more fully probe the nature of our humanity.

Afterword

We have reached the end of this book, but this is hardly the end of possibilities for drawing physics and dance together. Like a dance by Merce Cunningham, the science-art inquiry can continue in limitless combinations, independent of the curtain rising and falling on any single performance. For in reality no single standardized method of pulling art and science together exists. Instead, the exchange is specific to each project and to who is in the room. Equipped with tools from both disciplines, you, the inquirer, can shape the questions and the methods by which you seek answers. You can also determine the outcomes: will your research result in a calculation? Or a choreographic work? Perhaps both at once? The goal is to seek mutual illumination: points of contact in which both disciplines are productively viewed in new ways through the exchange.

Seeking new ways of seeing is a natural extension of our own backgrounds: we both grew up professionally within institutions created to foster innovation in the mid-twentieth century. While studying physics at Harvard University and the University of Rochester, Sarah Demers's research relied on the TeVatron at Fermilab, founded in 1967 to foster experimentation in high-energy physics. She later became a postdoctoral research scholar for Stanford University's Linear Accelerator Center and joined the ATLAS experiment at CERN, the European Organization for Nuclear Research established in 1954 to explore the fundamental particles that constitute our universe. She joined the Yale faculty in 2009, and her particle physics research continues through her membership in the ATLAS collaboration at CERN and the Mu2e collaboration at Fermilab.

After studying at the School of American Ballet, Emily Coates began her career as a member of New York City Ballet, established in 1948 to nurture George Balanchine's formidable choreographic talents. Jerome Robbins became associate artistic director in 1949, an affiliation that continued for five more decades, enabling her to work closely with Robbins at the end of his career. Through dancing in Balanchine's and Robbins's ballets, she learned that an aesthetic inheritance such as classical ballet can and should be altered in the hands of contemporary artists. She later performed in the companies of Mikhail Baryshnikov, Twyla Tharp, and Yvonne Rainer, artists whose engagement with dance history and virtuosic alterations of the art form continue to influence her choreographic work and teaching. As a faculty member at Yale, she created Yale's dance studies curriculum as an intertwining of her Yale education—in which she studied English literature and culture, history, and politics in American studies—and the embodied knowledge of her professional career.

All this is to say, whereas Demers explains something as mundane as slipping on the ice through subatomic physics, Coates sees the world in terms of choreographic form. We both take the spirit of innovation in our influences and spin off in other directions by pulling our aesthetic and scientific knowledge together.

Our backgrounds inform our writing in other ways, too. By *dance*, we are primarily referring to concert dance—choreography presented before a public—and the examples in the book come predominantly, though not exclusively, from prominent Western ballet, modern, and postmodern dance choreographers. We refer frequently to experimental or avant-garde artists, because they tend to be the ones who question the form and medium of dance most. There are many more great dance artists beyond those we had space for in these pages. Likewise, on the physics side, we had to narrow our scope and divide subjects in such a way as to create the dialogue with dance. In a book solely about physics, the topics would be sequenced and explained differently. Some precursors to our work are several books by Kenneth Laws, who expertly analyzes classical ballet through classical mechanics. Our work casts a wider net, by opening the lens to diverse dance forms, choreographic practices, and concepts in modern physics.

There is a point at which the productiveness of conjoining physics and dance breaks down. In simplifying bodies for the sake of analysis, physics cannot account for the politics of spectatorship: the ways that gender, race, sexuality, and class inform our reception of the human body in motion. Nor can it articulate why people dance, a joy for many that intensifies for some into political urgency. And while dance has much to offer a discussion of physics, the human body does have limits to what it can perceive. To access certain phenomena, nothing can replace the precision of a mathematical description or the extension of the human senses through scientific instruments.

Still, we have everything to gain by taking each other seriously as research partners. Consider Pina Bausch's dramatic falls through gravitational potential energy, or Einstein's $E = mc^2$ through Ralph Lemon's breakdown of choreographic form, and you can begin to understand and feel the concept of movement anew.

Early on, we developed a "Manifesto for Physics and Dance" to guide our collaborative work. The first line has become our mantra: "Physics and dance share equal creative, rigorous, intellectual research power." Armed with this tenet, go forth.

Acknowledgments

We developed many of the ideas in this book while co-teaching a course at Yale called the Physics of Dance. We thank Bill Segraves, then dean of science education, for coming up with the idea eight years ago for a cross-disciplinary course, and Susan Cahan, then dean of the arts, for helping him to bring us together. Little did we all know then where this connection would lead.

We are deeply grateful to the eighty-five students who have taken our class in the four times we have taught it since 2011. Their adventurous intellects and artistry have inspired our interdisciplinary inquiry. Our teaching fellow, Mariel Pettee, a talented particle physicist and dancer, proved to the class that one body could in fact hold the knowledge of both disciplines.

A number of colleagues within the university have offered invaluable support for our work, including current and former department chairs Meg Urry, Paul Tipton, Marc Robinson, and Daniel Harrison, and our colleagues within the Department of Physics and Theater Studies. Gary Tomlinson, Mark Bauer, and Norma Thompson at the Whitney Humanities Center, and the Franke Program in Science and the Humanities have offered us important platforms. Conversations with Mark Aronson, Paola Bertucci, Lacina Coulibaly, Stephen Davis, Miraj Desai, Kathryn Dudley, Jennifer Frederick, Steven Girvin, Inderpal Grewal, Iréne Hultman, Edward Kairiss, Kaury Kucera, Andrew Miranker, Priya Natarajan, Nikhil Padmanabhan, Renee Robinson, Brian Seibert, Laura Wexler, our Interdisciplinary Science and Art Research working group, and guests Jonathan Butterworth, Elizabeth Johnson, Rasika Khanna, Young-Kee Kim, Daniel Lepkoff, Liz Lerman, Brian Stewart, and Reggie Wilson have opened up new directions in our thinking. Richard Prum, ornithologist extraordinaire, has been a kindred interlocutor and supporter throughout our collaboration.

A number of individuals and institutions have offered us opportunities to expand our work into the public sphere: the Yale Alumni Association; Yale TEDx; Pierson College; Margaret Clark and Trumbull College; Cindy Clair and the Arts Council of Greater New Haven; the International Festival of Arts and Ideas; Joy Kasson at the University of North Carolina at Chapel Hill and Carolina Arts; Douglas Crimp and the University of Rochester; Brent Hayes Edwards and Columbia University's Heyman Center for the Humanities, with Eileen Gillooly, Marcia Sells, Pamela Smith, and the Embodied Cognition Working Group; Pamela Tatge at Wesleyan University; the Physics Club hosted by Yale's Department of Physics; and the physicists on ATLAS at CERN, especially Steve Goldfarb in public outreach. We are grateful to the leaders and staff of Danspace Project, Works & Process at the Guggenheim, the Wadsworth Atheneum, and the Center for Ballet and the Arts, which presented excerpts or all of the performance Emily Coates created based in part on our work. Jenai Cutcher, Johannes DeYoung, Jon Kinzel, and Liz Diamond have been instrumental to specific projects, and Yvonne Rainer has worn many different hats, as our guest speaker, performer, reader, adviser, and champion. Particle physicists Melissa Franklin, Mike Hildreth, Adam Martin, Toyoko Orimoto, Stephen Sekula, Mike Tuts, and Daniel Whiteson generously participated in our film, *Three Views of the Higgs and Dance* (2013).

Transforming our ideas into written form has added an entirely new dimension to our collaboration. Among those who have given us crucial support in that process, we thank foremost at Yale University Press Senior Editor Joseph Calamia for proposing that we write a book in the first place. His wisdom, good humor, and ongoing faith in our work keeps us on track. We also thank Senior Executive Editor Jean Thomson Black for her early advocacy of the project; Eva Skewes for assisting with the publication details; Sonia Shannon for her work on the jacket design; and Susan Laity, whose copyediting rivals the artful level of Balanchine's choreographic craft. The French horn player and composer Will Orzo offered us invaluable (and often hilarious) feedback on earlier drafts, pressing us to clarify our thoughts and simplify our language. Thomas DeFrantz at Duke University and Katie Glasner at Barnard College generously read through the manuscript, encouraging and challenging us with their comments—we are grateful for their expertise.

The movements and forces we describe in the book would be far less easy to envision without the photographs by Jessica Todd Harper and drawings by Eric Jiaju Lee. We thank both artists, as well as the Baryshnikov Arts Center for hosting our photoshoot in the incomparable natural light of the John Cage and Merce Cunningham Studio. We thank the dancers who enthusiastically joined us over those two days, made up of then-current students and alumni: Liam Appelson, Luna Beller-Tadiar, Derek DiMartini, Nicole Feng, Christina Kim, Suh Young Kim, Indrani Krishnan-Lukomski, Mariel Pettee, Elizabeth Quander, Holly Taylor, and Michaela Vitagliano. Created with ink and brush in the manner of Chinese calligraphy, Lee's drawings work in tandem with Harper's photographs to pull the essential forces described in the text into view. None of these images would have been created without the support of Doron Weber and a grant from the Alfred P. Sloan Foundation in the category of Public Understanding of Science, Technology, and Economics.

Our professional work is vastly enabled by the love of our parents, siblings, extended families, and close friends. At age 103, our oldest fan, Great-Aunt Mary Mimms, gave us the perspective of someone who saw modern physics enter the public discourse. Ramsey Coates's grace lives in this writing. We reserve special gratitude for our spouses, Steve Konezny and Will Orzo, who have supported our collaboration in countless ways—from childcare to cheerleading—with intelligence and generosity. Our children grew bigger (and one even arrived) as we worked on this book: we write to expand the possibilities for what they can imagine and thus become.

Workbook

Physics Problems

Gravity Exercises

1. The force an object feels due to gravity on earth is often referred to as the object's weight. This could be given in units of newtons (N) or force-pounds (lbs). Given that 1 N equals approximately 0.22 lbs, convert 500 N into lbs. Convert 150 lbs into N.

2. Calculate the magnitude of the gravitational force of attraction between two 75 kg dancers whose centers of mass are separated by 1 m. Then calculate the gravitational force of attraction that each dancer feels with the earth. What is the ratio between the two forces, dancer to dancer compared with dancer to earth? Assume that earth's mass is 6×10^{24} kg and that its radius is 6.4×10^6 m.

3. Your mass is a quantity that can change with time according to many factors. Your weight, in turn, depends on your mass and the magnitude and dimensions of the mass on which you are standing (mass and radius of the planet) and is the force that you calculate using Newton's Universal Law of Gravitation. If you were teleported to the moon, your mass would stay the same (assuming the teleportation didn't drop any body parts on the way), but your weight would change. If your mass is 80 kg, what would your weight be on the earth and on the moon? Report your answer in units of newtons and also in units of pounds. Assume that the earth's mass is 6×10^{24} kg, the earth's radius is 6.4×10^6 m, the moon's mass is 7.3×10^{22} kg, and the moon's radius is 1.7×10^6 m.

4. You have been sent, in a spacesuit and with a bathroom scale, to a new planet to determine its mass. If your total mass (you plus your spacesuit) is 120 kg, the planet's radius is 9×10^6 m, and on the planet's surface you weigh 400 lbs (with the spacesuit on), what is the mass of this newly discovered planet? (Hint: You need to first convert your weight in lbs to the equivalent force in newtons so that you are using consistent units throughout the problem.)

5. You introduce a choreographer to Newton's Universal Law of Gravitation and he or she immediately grasps the concept that each mass is gravitationally attracted to all other masses in the universe. The choreographer would like to stage a performance in the open space underneath the Eiffel Tower in order to increase the height and length of the leaps of the dancers because the gravitational pull of the Eiffel Tower will work against the gravitational attraction of the earth. The choreographer asks you to calculate the force due to the gravitational attraction between one of the 70 kg dancers and the Eiffel Tower, a force that would be acting in the direction of the sky. The Eiffel Tower has a mass of approximately 7.3 million kg. Its total height is over 300 m, but knowing that the mass is concentrated near the ground, you estimate the center of mass to be 80 m above the surface of the earth. Given these rough assumptions, calculate the gravitational force the dancer would experience due to the Eiffel Tower. How does that compare to the force the dancer feels due to proximity to the earth? Will the dancer be able to jump noticeably higher?

6. Three masses have been placed on the x-axis at the following positions. Calculate the center of mass of the system on the x-axis:
Mass 1: 50 kg, x-axis position at –3.5 m.

Mass 2: 80 kg, x-axis position at 0.0 m.

Mass 3: 75 kg, x-axis position at 4.0 m.

7. Draw an x-y coordinate system, indicate the location of the following two masses, and then calculate the center of mass of the combined system in both x and y. Mass 1 has a mass of 5 kg and is located at ($x = 1.0$ m, $y = -1.0$ m). Mass 2 has a mass of 2 kg and is located at ($x = 0.0$ m, $y = 1.0$ m).

8. Try balancing on one foot while reaching forward with your arms and upper body. Using what you learned in Chapter 1 about the conditions required for balancing, explain why it can be helpful to extend your leg in the opposite direction from your arms to maintain balance during this movement.

9. You are in a performance with three other dancers in which you all have been asked to climb up a four-sided tower structure that has been rolled out onstage. Each time you practice, the structure tips over as the dancers begin to climb. You point out to the director that this problem could be solved either by (a) having the dancers climb up the four sides of the structure simultaneously instead of all climbing up one side or by (b) having them all climb up the same side but first making the structure heavier. Use what you have learned about center of mass and the conditions for balance to justify each of these suggestions.

10. An 80 kg dancer is perfectly balanced en pointe, with her center of mass directly over the center of her shoe's contact with the floor. Assume that her pointe shoe's area of contact with the floor has a diameter of 5 cm and is a perfect circle. The dancer's left hand is extended such that it is 0.3 m in front of the vertical line that joins her center of mass with the floor. If someone places a 3 kg mass in her hand, can she maintain her balance without shifting her body? Remember that the condition for balance is that her center of mass lies over the area of support.

Force Exercises

11. Two skaters are on an ice rink. Skater A is wearing cleats that cut into the ice and provide enough traction so that she can apply a horizontal force to Skater B, without herself moving. Skater B wears slippery shoes and slides when pushed. In each of the following exercises, draw a free body diagram of Skater B and calculate his acceleration from Skater A's push.

   (a) Skater B has a mass of 55 kg and is pushed by Skater A with a force of 100 N.

   (b) Skater B has a mass of 55 kg and is pushed by Skater A with a force of 200 N.

   (c) Skater B has a mass of 110 kg and is pushed by Skater A with a force of 100 N.

   (d) Skater B has a mass of 110 kg and is pushed by Skater A with a force of 200 N.

12. For the same scenarios as problem 11(a)–11(d) draw a free body diagram of Skater A, taking into account the forces of gravity, the normal force from the

ice, friction due to the interaction between Skater A's cleats and the ice, and the equal and opposite force acting on Skater A due to contact with Skater B. Hint: Recall that Skater A is not moving, so the acceleration is 0.

13. Imagine a scenario in which one dancer (Dancer A) is being held in the air by another dancer (Dancer B). Draw a free body diagram of Dancer A. Then draw the corresponding free body diagram of Dancer B. Looking at your two diagrams side by side, identify where Newton's 3rd Law of Motion pairs of forces can be found. (These are the equal and opposite forces that exist any time surfaces are in contact.)

14. Draw a free body diagram of a dancer standing in the middle of a dance studio, not moving. Repeat this for the scenario in which the dancer is in the air performing a vertical jump. Will your free body diagram be different when the dancer is on the way up from when he or she is at the peak of the jump or on the way down?

15. Two dancers are standing on ice wearing slippery shoes. Imagine that the frictional forces between their shoes and the ice are so tiny that they can be ignored. Which of the following arguments is most correct if the two dancers push off against each other? Justify your answer using Newton's 2nd and 3rd Laws of Motion.

    (a) The dancer who is stronger will be the one who experiences a greater acceleration.

    (b) The dancer who is weaker will be the one who experiences a greater acceleration.

    (c) The dancer who has more mass will be the one who experiences a greater acceleration.

    (d) The dancer who has less mass will be the one who experiences a greater acceleration.

16. If a dancer pushes against the floor with a force of 450 N, what must the dancer's mass be if he or she achieves an acceleration of 5 m/s^2?

17. At the lowest point of a plié, a 50 kg dancer accelerates his center of mass upward by pushing against the floor. This acceleration occurs over a fraction of a second as he quickly reaches a constant velocity that he maintains until he is almost fully standing straight, at which point a deceleration occurs. In the following problems, calculate the force for a specific series of movements within a plié.

    (a) As the dancer pushes off the floor, he has a constant acceleration for 0.25 s at a rate of 2 m/s^2. What force must he apply to the floor, above the force of his weight, while he achieves this acceleration?

    (b) Near the top of the plié, there is an acceleration, this time at the rate of –2 m/s^2 over a period of 0.25 s. If you were measuring the force between the dancer's feet and the floor during this deceleration, what would you measure? Draw a force diagram for this scenario to help you calculate the net force acting on the dancer at this time.

    (c) Plot the force between the dancer and the floor as a function of time, beginning with the acceleration of 2 m/s^2 as described in part (a), followed by a constant velocity for half a second, followed by the acceleration of –2 m/s^2 as described in part (b).

18. In an assisted lift, Dancer A (with a mass of 60 kg) jumps and Dancer B (who has a mass of 80 kg) provides support throughout the ascent and descent. Throughout the entire jump the dancers remain in contact.

   (a) What will the magnitude of the force between Dancer B and the floor equal when Dancer B is applying an upward force of 75 N on Dancer A?

   (b) If there is a moment when Dancer B fully supports Dancer A in the air and Dancer A is not accelerating, what will the magnitude of the force between Dancer B and the floor equal?

   (c) Repeat parts (a) and (b) but switch the dancers' masses: Dancer A now has a mass of 80 kg and Dancer B has a mass of 60 kg.

19. A standing dancer goes to lean against what she assumes is a wall but is actually a movable barrier. When she makes contact with the barrier, it moves, and the dancer falls. Is Newton's 3rd Law of Motion broken in this instance? Describe the forces between the dancer and the movable barrier, and the forces between the movable barrier and the floor during the dancer's fall.

   A few minutes later, this very unlucky dancer is moving very quickly toward the real wall of the dance studio and hits it with such great force that the plaster on the wall becomes indented. Is Newton's 3rd Law broken in this instance? Describe the forces between the dancer and the wall as the plaster is breaking.

20. Using Newton's Laws:

   (a) Justify using a wooden floor over a concrete floor for a dance studio.

   (b) Describe some of the challenges of executing choreography that requires rapid acceleration of the dancers in a studio with a spongy, foam floor.

Friction Exercises

21. Classify the following as examples of static friction, kinetic friction, or both, and justify your answer:

   (a) Socks and floor: Dancing in socks on a slippery floor.

   (b) Sneakers and rubber-coated floor: Running in sneakers on a rubbery surface.

   (c) Lift: Two dancers with firmly clasped hands engaged in a lift.

   (d) Shoes and ice: Running on ice in smooth-soled shoes.

22. What will the magnitude of the force due to kinetic friction F_k be between two objects with a normal force F_N of 100 N between them if the coefficient of kinetic friction μ_k of the material pair is

   (a) 0.4?

   (b) 0.8?

   (c) 1.0?

23. Two slippery materials have a coefficient of static friction μ_s of 0.1 when put together. If they have a normal force F_N of 40 N between them, what maximum force parallel to their contact can they withstand before they slip?

24. Using the example in Chapter 4 in which you were in rubber-soled sneakers standing on concrete and found that when you pushed off with a force of 600 N at an angle of 40 degrees with respect to the ground, your sneakers slipped because the parallel force you applied was greater than the maximum force available from static friction F_{smax}, answer the following questions:

    (a) Would you have gotten the same result if the total force you applied was 400 N instead of 600 N?

    (b) Would you still have slipped if the angle had been 45 degrees?

    (c) Would you still have slipped if the angle had been 50 degrees?

25. You are practicing the surfing exercise from Chapter 4 in your socks on a floor with which your socks have a μ_k of 0.2. If your force on the ground (weight) is 800 N:

    (a) What will be the force of kinetic friction F_k on you while you are mid-slide?

    (b) Does your answer to part (a) depend on whether you are balancing on one or two feet?

    (c) Does your answer to part (a) depend on how fast you are going?

26. You are sliding around a room with a polished wooden floor in your socks. The coefficient of kinetic friction between your socks and the floor is $\mu_k = 0.25$. Assuming your weight is 800 N, what will be the force of kinetic friction F_k on you while you are mid-slide?

    Now someone walks by and hands you a box that weighs 150 N. When you slide carrying the box, what value of F_k will you experience?

27. You are in a performance in which you stand on a platform that tilts during the show. Initially it is horizontal. You remain standing on the platform, but at some angle of inclination of the platform your feet begin to slide. An audience member approaches you after the program wanting to know how you started to spontaneously move when you had been standing completely still and did not lift your feet. Use the language of the forces of friction, including the normal force F_N, to explain how it happened. Include diagrams with your answer.

28. In locations with many winter storms and steep terrain people may take the following precautions. Explain each modification below using arguments related to the equations for friction that you have learned.

    (a) People put chains around their automobile tires.

    (b) People put bags of dirt in the trunks of their cars.

    (c) People make modifications to their wardrobe or gear when they go out hiking up a steep ice bank. List some example modifications in your explanation.

29. You are in a performance where you need to pull an object along the floor. You have been rehearsing in bare feet, and all has gone well, though the object occasionally gets stuck while you are dragging it. The costume designer has been told that there will be sliding in the performance and that it hasn't been going as well as it should. The designer brings socks for you to wear in

the performance. Write an explanation to the designer of why the socks will make the slipping worse and clarify what you need, using arguments related to kinetic and static friction.

30. Why do we have one value for the force due to kinetic friction F_k, but a maximum value for the force due to static friction F_s?

Motion Exercises

31. Draw a free body diagram of yourself doing this problem set. If you are resting your hands on some object, sitting, or leaning against something, be sure to take into account all of the resulting forces acting on you.

32. Draw a free body diagram of a dancer in the following scenarios that include a net force and therefore a net acceleration. If there is more than one force acting on the dancer, note which one must have a greater magnitude.

    (a) The dancer is on the ground, an instant before launching into a jump from one foot.

    (b) The dancer is in mid-air, after taking off from the jump in part (a).

33. A dancer known for expertly landing after jumping down from various objects has stipulated that he or she must not land on the ground with a velocity higher than 6 m/s in a given performance. What is the maximum height from which the dancer is willing to step off to fall?

34. You step off a box that is 0.5 m off the ground. Assuming that you have no initial velocity in the vertical (y) direction, how fast will you be going by the time your feet hit the floor? How long with the drop take? (Assume that you do not begin to bend your knees until your feet make contact with the ground.)

35. A choreographer is working with a musical score and trying to line up dancers to jump to the music. Answer the following questions, keeping in mind that in a total jump time, the dancer will spend half of that time going up and half falling back down.

    (a) In a first attempt, the choreographer asks for each dancer to be in the air for a total of 0.2 s. How high would the dancer need to jump? How fast must the dancer be going in the +y direction when leaving the ground?

    (b) The choreographer now uses a different piece of music, with a much slower tempo. The dancers are asked to be in the air for 1 s to accommodate the new music. How high would the dancers need to jump in order to be in the air for 1 s?

36. Two dancers of equal strength are attempting to leap as far across the stage as they can in one jump. One of the dancers begins from rest and the other gets a running start. Using the equations of projectile motion, make an argument for why the dancer with a running start will travel farther, even if the two are in the air for the same period of time.

37. A dancer who wants to jump higher decides that a way to do it is to get more force on the ground. To get this increased force the dancer holds a number of heavy rocks while jumping. Using your physics and dance knowledge, explain to the dancer why this technique did not result in higher jumps.

172

38. Describe two methods that you could use to measure the maximum height that you are able to jump. (Try to rely on only material/equipment that you have on hand.) Discuss the weaknesses of each of your methods both from the perspective of variability of your motions and the precision of the required measurements.

39. Carry out one of the experiments that you designed for the previous question. Include a description and diagram of your set-up, your full data set, and your resulting measurement with an estimated uncertainty.

40. Using Newton's laws and the equations of projectile motion, explain why you are able to jump higher when jumping on a trampoline than when you jump on the ground.

Momemtum Exercises

41. A dancer with a mass of 75 kg moves along a straight line at 2.5 m/s.

    (a) What is the magnitude of the dancer's momentum?

    (b) The dancer wants to move across the room again, but this time with double the momentum. A friend hands the dancer a 20 kg object to carry to increase their momentum. What speed will the dancer need to move in order to double their momentum from part (a) while carrying the 20 kg object?

42. You are traveling with a velocity of 3 m/s in a direction that you have designated as along the positive x-axis in your dance studio. If a dancer with twice your mass is moving through the studio in the opposite direction, how fast should the dancer be moving in order to have a magnitude of momentum equal to yours?

43. You are moving quickly through the studio and attempt to stop on a dime, going from full momentum to zero momentum in an instant, but while you manage to stop your feet, your body topples over in the direction you were moving. Explain why this happens.

44. You are in outer space holding a flashlight and attempting to get back to your spaceship by it. Assume that you have a mass of 80 kg (including the space suit!) and the flashlight has a mass of 1 kg. If you throw the flashlight directly away from the ship with a speed of 10 m/s:

    (a) What is the momentum of the flashlight after the throw?

    (b) What is your momentum (including the space suit) after the throw?

    (c) With what speed will you be moving toward the ship?

    (d) If you need to travel a distance of 10 m, how long will it take?

45. Repeat the previous problem, but assume that you have thrown the flashlight with a speed of 25 m/s.

46. You are sitting in the middle of an iced-over pond and need to get to the edge, but the ice is so slippery that you cannot get traction with your shoes. You have an apple in your backpack that you were going to save for an afternoon snack, but you realize that instead you can use it as a partner in a conservation

of momentum dance to project yourself toward the edge of the pond. If your mass is 70 kg and the apple has a mass of 0.2 kg:

(a) how fast must you throw the apple in order to get yourself moving at a speed of 0.25 m/s?

(b) Convert this speed to miles per hour. Do you think your calculated speed is achievable, or will you need to follow your apple toss with throwing something else in your backpack? (Note that a major league baseball pitcher can throw a fastball at about 90 mph.)

47. A 100 kg dancer is standing on a slippery ice rink next to an 80 kg dancer. The two push off against each other and begin moving apart on the ice. Assume that there is no friction between the dancers and the ice, so momentum will be conserved. Take an instant where the two of them are initially at rest as the initial snapshot and a moment when they have lost contact and are moving apart as the final snapshot for the following momentum conservation problem. Include a diagram with your answer.

(a) What is the initial momentum of the system that includes both dancers?

(b) What will the final momentum be of the system of both dancers?

(c) If the 100 kg dancer ends up moving with a speed of 4 m/s, at what speed will the 80 kg dancer be moving?

(d) What is the momentum of each individual dancer (giving both magnitude and direction) after the two push off? (Be sure that the direction you quote in your answer matches your diagram.)

48. Two dancers move toward each other on the ice and are expected to clasp their arms together and remain stationary where they meet. Dancer A has a mass of 60 kg and Dancer B has a mass of 75 kg. They initially try moving toward each other at the same speed, but when they clasp arms they drift off along the direction of one of the dancers. You point out that they have different masses and so their final motion together is simply conserving momentum. If Dancer A moves toward Dancer B with a speed of 1 m/s, how fast should Dancer B be moving toward Dancer A if the two want to end up at rest?

49. You are involved with a performance on a large trampoline during which you will spend a lot of time in the air and will need to navigate in-air collisions with other dancers. While you are in the air, the only external force acting on the group of dancers is due to gravity, so you can assume that your collective horizontal momentum will be conserved. If two dancers with mass of 60 kg each are moving through the air at 0.5 m/s, how fast should an 80 kg dancer be moving in the opposite direction if the three of them want to have zero horizontal momentum after colliding and ending together in a heap?

50. A choreographer has requested a group of dancers to perform on ice, but the budget for the performance was not sufficient to hire a rink. They therefore need to move as if they were on ice, while actually being on a wooden stage floor. What instructions would you give the dancers in terms of how to interact with each other and the floor in order to make it seem as if they are on ice? Use the ideas of conservation of momentum in your instructions and include guidelines for individuals moving alone and for times when dancers come into contact with each other.

Turning Exercises

51. If you apply twice the torque to an object without changing its moment of inertia, will the resulting initial angular acceleration increase, decrease, or stay the same? If the angular acceleration changes, by how much does it change?

52. If you apply twice the torque to an object and at the same time double the moment of inertia of the object, will the resulting initial angular acceleration increase, decrease, or stay the same? If the angular acceleration changes, by how much does it change?

53. You need to loosen a nut with a wrench that is 25 cm in length. To do this you apply a 150 N force to the end of the wrench.

    (a) If you apply the force at the end of the wrench at an angle of 90 degrees with respect to the line defined by the length of the wrench, what torque have you applied?

    (b) If, instead, you apply a force at an angle of 45 degrees with respect to the line defined by the length of the wrench, what torque have you applied?

    (c) What force would you need to apply at 45 degrees in order to result in the torque calculated in part (a), where the force of 150 N was applied at 90 degrees?

54. You want to lift a heavy boulder using a wooden plank and triangular wedge to create a lever.

    (a) In order to maximize the force due to torque that you apply to lift the boulder, would you place the wedge closer to the boulder or closer to your hands? Defend your answer using the formula for torque defined in Chapter 6.

    (b) If the plank is 1.5 m long, the boulder has a mass of 60 kg, and the wedge is placed 0.5 m from the boulder, what force would you need to apply to the other end of the plank (1.0 m from the wedge) to equal the force due to gravity? Assume that the boulder is placed on a platform attached to the end of the plank such that the entire force due to torque is transferred to the vertical direction to oppose gravity.

55. Consider a system of three 1 kg masses placed in the plane formed by the x-y axes. The masses are placed at locations (1.0 m, 1.0 m), (0.0 m, 1.0 m), and (1.0 m, 1.0 m) in (x,y).

    (a) What is the moment of inertia of the system of these masses about the axis of rotation defined by the z-axis passing through the point (0.0 m, 0.0 m) in the x-y plane?

    (b) Would adding a 1 kg mass to the point (0.0 m, 0.0 m) in the x-y plane change the moment of inertia about this axis of rotation? If yes, by how much? If no, why not?

56. Using the technique documented in Chapter 6, calculate the moment of inertia for the right arm of a 60 kg woman with average dimensions as shown in the table in the starting position of the classical Russian pirouette. Assume, as in the diagram in the text, that the forearm makes a right angle with the upper arm. Compare your result with the result calculated in the text for the Balanchine technique.

57. If you hold your body in a plank position (standing up straight and rigid) and tip forward until you start to fall, the force due to gravity that acts on your center of mass can be thought of as a torque on your body that causes it to rotate.

   (a) If you assume that your mass is 75 kg, you are 2 m tall, and your center of mass is exactly halfway between the bottom of your feet and the tip of your head, what would the torque due to gravity be on your body when your body is at a 5-degree angle from the vertical?

   (b) What is the torque due to gravity when your body is at a 10-degree angle from the vertical?

58. You are standing on a turntable at a park that one of your friends has pushed to give an initial rotation. Neglect friction in the spinning of the table.

   (a) If you start to feel sick due to the spinning motion of the table and you want to slow it down, would you walk toward the center of the table, which is the point about which it is rotating, or would you walk toward the edge? Defend your answer in terms of the quantities defined in Chapter 6.

   (b) After a bit of time at the slower rotational speed you recover and decide that you want the spinning rate to increase. In what direction would you walk to speed up the rotation? At what point on the turntable would you stand to maximize the rate of rotation? Again, defend your answer using quantities defined in Chapter 6.

59. A common technique in ice skating involves changing the speed of rotation in a turn by moving limbs toward and away from the axis of rotation. In order to get a sense of the quantitative difference between a position with limbs extended and one with the limbs drawn into the body, calculate the change of the moment of inertia for a 50 kg person whose leg is extended from the body at a 90-degree angle with the vertical compared to when the leg is pulled in to the body, parallel with the vertical. In your calculation assume that the person spins on the leg that is not being extended and the extended leg is 20% of the total body weight of the person. Describe all further assumptions that you make in your calculation.

60. Using the physics principles of turning that we have described in Chapter 6, design an efficient and powerful turn, justifying the choices you have made with the language of torque, moments of inertia, and angular momentum. Note that your axis of rotation does not need to pass through one of your feet, so you can think of rotations of your body in configurations other than a pirouette.

Choreographic Studies

The choreographic studies that follow transform physics information into creative tools to generate movement. Most of the studies are based on the assumption that you will be working in a group. Each study is intended to enhance your understanding of the physics involved, while also guiding you through a choreographic method. The method we teach here is commonly used in Western postmodern and contemporary dance and involves developing movement phrases, which are sequences of movements that contain spatial and temporal information. These phrases are manipulated and assembled to create the larger work. A phrase is a known with which to venture into the unknown of your choreographic research.

The process we lead you through has three phases: generating movement phrases from different sources; researching that material using choreographic manipulations of energy, time, and space; and composing a dance, which can take many different forms, with the resulting material.

Moving from researching to composing introduces new considerations, such as sequencing, transitions, compositional arc, and overall rhythmic composition. Context—the location of the phrase within a dance—affects reception. The relationship of the viewer to the dance is another consideration.

What is the physics of the final composition? How much of the physics you started with should be apparent in the final dance?

At a certain point, choreographic research should take over. You should not feel required to "illustrate" the physics but instead allow yourself to develop the compelling qualities that you observe in the movement. What those qualities are can vary widely. You must learn to see like a choreographer.

The residue of the physics that launched you will remain in some form in the final work. The physics may not be literally apparent (to anyone who has not also read this book), but it will still be present. You will be thinking about the physics in a new medium. The ideas will take on other resonances.

A question more generative to this process is what can you as the creator take from paying attention to physics? Look for choreographic scores; ideas about energy, space, and time to apply to movement material; anything that strikes you as worth experimenting with. Physics changes, in this process, into a resource of ideas that make you want to move and inspire you to create. As the author, you are in charge of the information.

Some dangers in dance-science work include excessive literalism, cliché, and lack of attention to the science. But context is everything, and some of these dangers might actually lead to interesting results if framed well.

Developing finely tuned skills of observation is essential both in choreographic research and in physics.

These choreographic studies are intended to give you ideas. You can always design your own.

1. **Cartesian Coordinates: X, Y, and Z**

   Create a movement phrase that manipulates center of mass. The phrase must consist of five linked positions that satisfy these specifications:

   Position #1: Select a position, and identify your approximate center of mass on the x-, y-, and z-axes.

Position #2: Change your center of mass along the x-axis, keeping y and z center of mass constant.
Position #3: Preserve your center of mass on the x- and y-axes and change z.
Position #4: Change your y and z center of mass and preserve x.
Position #5: Restore to position #1.

Develop creative ways of shifting your limbs in order to alter or preserve your center of mass along the given axes. (Remember that the various parts of your body can move, not only your legs and torso!) Once you have identified your five positions, connect them kinesthetically. You may be surprised to learn how many options there are for getting from A to B. Be sure to clarify what the transitions will be between the positions.

2. **Speed, Acceleration, and Velocity**

   Manipulate your five-position Cartesian-coordinate phrase six different ways, according to these criteria:

	Velocity	Acceleration
1	0	0
2	+	0
3	+	+
4	+	-
5	-	0
6	-	-

3. **Statics and Dynamics**

   This study should be done in a group of two or more. Create a phrase consisting of six positions according to the following instructions:

   - each member of the group should develop a different phrase
   - you may choose to share some of the same positions within your phrases but not all
   - four out of the six positions should be stable
   - two out of the six positions should be unstable
   - two out of the four positions that are static/stable should require some kind of leaning support on your fellow dancers (consider hips, heads, feet, shoulders, and thighs—hands are not the only body part available for support)
   - come up with a way to sequence through these poses seamlessly as a group
   - look for a formation for the group as a whole that maximizes the progression of your poses in terms of design or mechanics, or both.

4. Momentum

Design a simple repetitive phrase that includes at least two changes in momentum. The key is to create a phrase that repeats. Create different scores for each member of the group, based on manipulations of velocity within the phrase. (For example: Dancer A moves forward at v for 2 seconds, then in reverse at v for 2 seconds; Dancer B moves forward at v for 1 second, then in reverse at v for 3 seconds, and so on). Practice executing the phrase with these varied scores simultaneously.

5. Turning

Design a new turn, using any orientation of your body in relation to the floor. Define the axis of rotation and draw a sketch of the turner in mid-turn. Consider the force that makes the turn begin (torque) and the resistance in the turn in terms of both moment of inertia and friction.

6. Energy

(a) Reconfigure the phrase you created in study #1 using manipulations of energy. Inject the movements with different energies: you might mix floating and vibrating qualities, or try adding a moment of stillness somewhere. Remember that you have available to you not only a full spectrum of movement qualities but also an understanding of the physics concepts and their formulas, which you may choose to manipulate. These include the notions of gravitational potential energy, spring potential energy, and kinetic energy.

(b) Using this same movement phrase, design a duet that uses the concept of the "energy response," in which you set the material into conversation with another dancer. You may choose to mobilize contrasting or corresponding energies, or a mix of both, with your partner. The important thing is to think about how a change in one element affects the others. This study should be 1.5 to 2 minutes long.

7. Space

"Iron out" one of your movement phrases by advancing it along a line.

Pairing up, execute your newly configured phrase alongside that of another dancer.

8. Time

Select one of your movement phrases. Create five new phrases by manipulating the internal transitions between movements or positions in the following ways:

- Continuous: Move as smoothly as possible through the phrase. Erase the divisions between discrete positions.

- Sharp: Break up the phrase into discrete pieces by striking and holding positions as you progress through the sequence. You may choose to pause the action in the full arrival into any given pose or while caught in between.

- Quick-to-slow: Move quickly into each position or picture and hold for a couple beats before slowly letting that position dissolve. Once the image has fully dissolved, move quickly into the next position.

- Slow-to-quick: Move slowly through the transitions between positions, speeding up just as you arrive in each position.

- Gooey: Erase the outline of the image and perform only its gooey interior.

Now perform these variations to pieces of music from different genres. Hold on to the integrity of your altered phrases, while allowing the music to inflect your movement quality.

9. **Final Project**

Create a choreographic composition that explores the perceptual shift between classical and modern physics.

Begin by choosing two concepts—one from classical physics and the other from modern physics—to research in greater depth. You may draw from an array of source material to deepen your understanding, including textbooks, journalism, scholarly articles and reports on advancements in the field, and books written on the topic for laypersons. Develop a bibliography.

Next, get moving! Develop movement phrases derived in some way from your research. Consider the mathematical formulas and their implications as useful points of departure for your creative investigation. Search for choreographic scores, or organizations of movement, that you can use to inform your choreographic material.

Compose these phrases into a choreographic composition. Craft a beginning, a middle, and an end, and set the material within a performance space. Consider the placement of the viewers. Draw on some of the choreographic strategies covered in this book and any others that you know or might invent.

Notes

[1] Lacina Coulibaly, "Sigini: Study of Movement (Foundation and Efficiency)" (in Emily Coates's possession, May 14, 2018).

[2] Isadora Duncan, "The Dance of the Future," in *Dance as a Theatre Art: Source Readings in Dance History from 1581 to the Present*, ed. Selma Jeanne Cohen (Princeton, NJ: Princeton Book, 1992), 124.

[3] Ann Daly, *Done Into Dance: Isadora Duncan in America* (Middletown, CT: Wesleyan University Press, 2002), 12.

[4] Anthea Kraut, "Between Primitivism and Diaspora: The Dance Performances of Josephine Baker, Zora Neale Hurston, and Katherine Dunham," *Theater Journal* 55, no. 3 (October 2003), 433-450.

[5] Kariamu Welsh-Asante, "In Memory of Pearl Primus," in *African Dance: An Artistic, Historical, and Philosophical Inquiry*, ed. Kariamu Welsh-Asante (Trenton, NJ: Africa World Press, 1996), x.

[6] Pearl Primus, "African Dance," in *African Dance*, 6–7.

[7] Pearl Primus, *Spirituals*, Jacobs Pillow Interactive, accessed November 4, 2017, https://danceinteractive.jacobspillow.org/pearl-primus/spirituals/.

[8] "From MR's Archives: Yvonne Rainer and Aileen Passloff in Conversation with Wendy Perron," Critical Correspondence, Movement Research, accessed November 4, 2017, http://old.movementresearch.org/criticalcorrespondence/blog/?p=10835.

[9] *Serway Physics for Scientists and Engineers*, 4th ed. (Orlando, FL: Harcourt Brace College Publishers, 1994), 126.

[10] For an in-depth study of contact improvisation, see Cynthia Jean Novak, *Sharing the Dance: Contact Improvisation and American Culture* (Madison, WI: University of Wisconsin Press, 1990).

[11] *Chute* (1979), *Videoda Contact Improvisation Archive: Collected Edition, 1972–1983*, DVD.

[12] *World's Most Talented*, W Channel, published on YouTube April 15, 2015, accessed May 29, 2018, https://www.youtube.com/watch?v=EZfVAxG2-h4.

[13] Elizabeth Kendall, *Balanchine and the Lost Muse: Revolution and the Making of a Choreographer* (New York, Oxford University Press: 2013), 43–44.

[14] Suzanne Farrell, with Toni Bentley, *Holding On To the Air: An Autobiography* (New York: Simon and Schuster, 1990), 94–95.

[15] Balanchine was not the first to teach the straight back leg in fourth position. See Fernau Hall, *Olga Preobrazhenskaya: A Portrait* (New York: M. Dekker, 1978), 134. Elizabeth Kendall, email to Emily Coates, May 30, 2018. His innovation was the long, low, deep lunge.

[16] Robert Enoka, *The Neuromechanics of Human Movement* (Champaign, IL: Human Kinetics Publishers, 2015), Table 2.5.

[17] *Tordre*, choreographed by Rachid Ouramdane, Baryshnikov Arts Center, October 14, 2016.

[18] Katherine Profeta, *Dramaturgy in Motion: At Work on Dance and Movement* (Madison, WI: University of Wisconsin Press, 2015), 40.

[19] Ibid., 81.

[20] Royona Mitra, "Akram Khan: Dance as Resistance," *Seminar*, December 6, 2015, accessed January 30, 2018, http://www.akramkhancompany.net/wp-content/uploads/2015/12/Royona-Why-Dance-piece.pdf.

[21] Susan Foster, "Dancing Bodies: An Addendum, 2009," *Theater* 40, no. 1 (2010), 27.

[22] Lynnette Young Overby and Jan Dunn, "The History and Research of Dance Imagery: Implications for Teachers," *The IADMS Bulletin for Teachers* 3, no. 2 (2011), 9.

[23] Deborah Friedes Galili, "Gaga: Moving Beyond Technique with Ohad Naharin in the 21st Century," *Dance Chronicle* 38, no. 3 (2015), 370.

[24] Yvonne Rainer, "A Quasi Survey of Some 'Minimalist' Tendencies in the Quantitatively Minimal Dance Activity Amidst the Plethora, or an Analysis of *Trio A*" in *Rainer, A Woman Who: Essays, Interviews, Scripts* (Baltimore, MD: Johns Hopkins University Press, 1999), 33.

[25] Joellen A. Meglin, Jennifer L. Conley, and Dakin Hart, "Ruth Page and Isamu Noguchi's *Expanding Universe* (1932, 1950, 2017)," Lecture Demonstration, Dance Studies Association Inaugural Conference, Columbus, OH, October 21, 2017.

[26] Deborah Hay, *Using the Sky: A Dance* (New York: Routledge, 2016), 8, 14.

[27] Ibid., 15.

[28] *Deborah Hay, not as Deborah Hay: A Documentary*, by Ellen Bromberg (2011), accessed November 11, 2017, https://vimeo.com/36519099.

[29] Hay, *Using the Sky*, 4.

[30] Edward Neville Da Costa Andrade, *Rutherford and the Nature of the Atom* (Garden City, NY: Doubleday, 1964) , 111.

[31] James Mooney, "Preface," *The Ghost Dance Religion and the Sioux Outbreak of 1890* (Lincoln: University of Nebraska Press, 1991), xxi; Alex K. Carroll, M. Nieves Zedeño, and Richard W. Stoffle, "Landscapes of the Ghost Dance: A Cartography of Numic Ritual," *Journal of Archaeological Method and Theory* 11, no. 2 (June 2004), 137.

[32] Mooney, "Preface," x.

[33] A. L. Kroeber, "A Ghost Dance in California," *Journal of American Folklore* 17, no. 64 (January–March 1904), 32–35.

[34] Carroll, Zedeño, and Stoffle, "Landscapes," 143–144.

[35] Carroll, Zedeño, and Stoffle, "Landscapes," 140–141.

[36] Mooney, "Preface," xvi; Russell Thornton, *We Shall Live Again: The 1870s and 1890s Ghost Dance Movements as Demographic Revitalization* (New York: Cambridge University Press, 1986), 12–13.

[37] Carroll, Zedeño, and Stoffle, "Landscapes," 129.

[38] Brenda Farnell, "Movement Notation Systems," in *The World's Writing Systems*, ed. Peter T. Daniels and William Bright (New York: Oxford University Press, 1996), 866.

[39] Ibid., 855.

[40] Ibid., 858.

[41] Rudolf Laban, *Choreutics*, ed. Lisa Ullman (Alton, Hampshire, UK: Dance Books, 2011), 4.

[42] Ibid., 10–11.

[43] Ibid., 17.

[44] Marion Kant, "German Dance and Modernity: Don't Mention the Nazis," in *The Routledge Dance Studies Reader: 2nd ed.*, ed. Alexandra Carter and Janet O'Shea (New York: Routledge, 2010), 113.

[45] Rob Iliffe, ed., *Early Biographies of Isaac Newton, 1665–1880, vol. 1* (London: Pickering and Chatto, 2006), 258.

[46] "Room Writing," *William Forsythe Improvisation Technologies: A Tool for the Analytical Dance Eye*, directed by William Forsythe (Karlsruhe: ZKM, 1999).

[47] Doris Humphrey, *The Art of Making Dances* (New York: Rinehart, 1959), 80.

[48] Jonathan Burrows, *A Choreographer's Handbook* (New York: Routledge, 2010), 100.

[49] Brian Seibert, "Faustin Linyekula: Remember His Name (and Country and Past)," *New York Times*, September 5, 2017, accessed November 11, 2017.

[50] "Einstein's Spacetime," *Gravity Probe B: Testing Einstein's Universe*, accessed May 31, 2018, https://einstein.stanford.edu/SPACETIME/spacetime2.html.

[51] Merce Cunningham, "Space, Time and Dance," in *Merce Cunningham: Dancing in Space and Time: Essays, 1944–1992*, ed. Jack Anderson and Ric Kostelanetz (Chicago: A Cappella Books, 1992), 37.

[52] Ibid., 39.

[53] Ibid.

[54] Ibid.

[55] Carolyn Brown, *Chance and Circumstance: Twenty Years with Cage and Cunningham* (New York: Alfred A. Knopf, 2007), 40–41.

[56] Ibid., 40.

[57] Suzanne Carbonneau, "Naked: Eiko and Koma in Art and Life," in *Time Is Not Even, Space Is Not Empty*, ed. Joan Rothfuss (New York: D.A.P./Distributed Art Publishers, 2011), 19.

[58] Olga Viso, "Foreword," *Time Is Not Even, Space Is Not Empty*, 14.

[59] Eiko Otake, "Like a River, Time Is Naked," presented as part of a 24-Hour Program on the Concept of Time, Solomon R. Guggenheim Museum, New York City, January 7, 2009, Eiko & Koma website, accessed November 11, 2017, http://eikoandkoma.org/index.php?p=ek&id=1989.

[60] Paul Kaiser, "Steps," OpenEndedGroup website, accessed November 12, 2017, http://openendedgroup.com/writings/steps.html.

[61] Danielle Goldman, *I Want to Be Ready: Improvised Dance as a Practice of Freedom* (Ann Arbor, MI: University of Michigan Press, 2010), 124–125.

[62] Ann Dils, "The Ghost in the Machine: Merce Cunningham and Bill T. Jones," *Performing Arts Journal* 70 (2002), 101.

Index